Wireless Networks and Mobile Computing

Wireless Networks and Mobile Computing

Edited by **Rafael Barrett**

WILLFORD PRESS

New York

Published by Willford Press,
118-35 Queens Blvd., Suite 400,
Forest Hills, NY 11375, USA
www.willfordpress.com

Wireless Networks and Mobile Computing
Edited by Rafael Barrett

© 2016 Willford Press

International Standard Book Number: 978-1-68285-066-4 (Hardback)

This book contains information obtained from authentic and highly regarded sources. Copyright for all individual chapters remain with the respective authors as indicated. All chapters are published with permission under the Creative Commons Attribution License or equivalent. A wide variety of references are listed. Permission and sources are indicated; for detailed attributions, please refer to the permissions page and list of contributors. Reasonable efforts have been made to publish reliable data and information, but the authors, editors and publisher cannot assume any responsibility for the validity of all materials or the consequences of their use.

The publisher's policy is to use permanent paper from mills that operate a sustainable forestry policy. Furthermore, the publisher ensures that the text paper and cover boards used have met acceptable environmental accreditation standards.

Trademark Notice: Registered trademark of products or corporate names are used only for explanation and identification without intent to infringe.

Printed in the United States of America.

Contents

	Preface	**VII**
Chapter 1	**Smart Living Using Bluetooth-Based Android Smartphone** Ming Yan and Hao Shi	1
Chapter 2	**On The Support of Multimedia Applications Over Wireless Mesh Networks** Chemseddine Bemmoussat, Fedoua Didi and Mohamed Feham	9
Chapter 3	**Clone-Based Mobile Agent Itinerary Planning Using Separate Trees for Data Fusion in WSNs** Soheil Javadi, Mohammad H. Hajiesmaili, Behzad Moshiri and Ahmad Khonsari	26
Chapter 4	**Performance Evaluation of MAC Protocols for Ad-Hoc Networks Using Directional Antenna** Arvind Kumar and Rajeev Tripathi	42
Chapter 5	**On Using Multi Agent Systems in Cognitive Radio Networks: A Survey** Emna Trigui, Moez Esseghir and Leila Merghem_Boulahia	54
Chapter 6	**An Intercell Interference Coordination Scheme in LTE Downlink Networks Based on User Priority and Fuzzy Logic System** A. Daeinabi, K. Sandrasegaran and X.Zhu	70
Chapter 7	**Reliability of Mobile Agents for Reliable Service Discovery Protocol in MANET** Roshni Neogy Chandreyee and Chowdhury Sarmistha Neogy	86
Chapter 8	**Impact of Different Mobility Scenarios on FQM Framework for Supporting Multimedia Applications in MANETs** Mohammed Saghir	101
Chapter 9	**An Educational Bluetooth Quizzing Application in Android** Michael Hosein and Laura Bigram	116

Chapter 10	**A Novel Approach for Mobility Management in LTE Femtocells** Pantha Ghosal, Shouman Barua, Ramprasad Subramanian, Shiqi Xing and Kumbesan Sandrasegaran	**126**
Chapter 11	**Bandwidth Aware on Demand Multipath Routing in MANETs** Tripti Sharma and Dr. Vivek Kumar	**140**
Chapter 12	**Highly Reliable Multi-Service Provisioning Using Sequential Prediction of Zone and PL&T Of Nodes in Mobile Networks** Sharmistha Khan, Dr. Dhadesugoor R. Vaman and Siew T. Koay	**151**
Chapter 13	**Impact of Random Mobility Models on OLSR** P. S. Vinayagam	**168**
Chapter 14	**A Review on Cooperative Communication Protocols in Wireless World** Juhi Garg, Priyanka Mehta and Kapil Gupta	**182**

Permissions

List of Contributors

Preface

The world is advancing at a fast pace like never before. Therefore, the need is to keep up with the latest developments. This book was an idea that came to fruition when the specialists in the area realized the need to coordinate together and document essential themes in the subject. That's when I was requested to be the editor. Editing this book has been an honour as it brings together diverse authors researching on different streams of the field. The book collates essential materials contributed by veterans in the area which can be utilized by students and researchers alike.

Science and technology have progressed swiftly over the past two decades. The applications of wireless networks and mobile computing have grown exponentially, especially in the industrial sector. This book throws light on diverse aspects of wireless networks and mobile computing through lucid elucidation of topics like software, internet applications, networking tools, etc. The various studies that are constantly contributing towards advancing technologies and evolution of this field are examined in detail. This book is a complete source of knowledge on the present status of this discipline and aims to serve as a resource guide for students and experts alike while contributing to the progress of this field.

Each chapter is a sole-standing publication that reflects each author's interpretation. Thus, the book displays a multi-facetted picture of our current understanding of applications and diverse aspects of the field. I would like to thank the contributors of this book and my family for their endless support.

Editor

SMART LIVING USING BLUETOOTH-BASED ANDROID SMARTPHONE

Ming Yan and Hao Shi

College of Engineering and Science, VictoriaUniversity, Melbourne, Australia
ming.yan2@live.vu.edu.au and hao.shi@vu.edu.au

ABSTRACT

With the development of modern technology and Android Smartphone, Smart Living is gradually changing people's life. Bluetooth technology, which aims to exchange data wirelessly in a short distance using short-wavelength radio transmissions, is providing a necessary technology to create convenience, intelligence and controllability. In this paper, a new Smart Living system called home lighting control system using Bluetooth-based Android Smartphone is proposed and prototyped. First Smartphone, Smart Living and Bluetooth technology are reviewed. Second the system architecture, communication protocol and hardware design aredescribed. Then the design of a Bluetooth-based Smartphone application and the prototype are presented. It is shown that Android Smartphone can provide a platform to implement Bluetooth-based application for Smart Living.

KEYWORDS

Android smartphone, Smart Living, Bluetooth module, single chip microcomputer, home automation

1. INTRODUCTION

Nowadays, smartphones are becoming more powerful with reinforced processors, larger storage capabilities, richer entertainment functions and more communication methods. Bluetooth, which is mainly used for data exchange, add new features to smartphones. Bluetooth technology, created by telecom vendor Ericsson in 1994 [1], shows its advantage by integrating with smartphones. It has changed how people use digital devices at home or office, and has transferred traditional wired digital devices into wireless devices. A host Bluetooth device is capable of communicating with up to seven Bluetooth modules at the same time through one link [2]. Considering its normal working area of within eight meters, it is especially useful in a home environment. Thanks to Bluetooth technology and other similar techniques, the concept of Smart Living has offered better opportunity in convenience, comfort and security which includes centralized control of air conditioning, , lighting, heating and cooling at home, and service robots[3] [4]. With dramatic increase in smartphone users, smartphones have gradually turned into an all-purpose portable device and provided people for their daily use [5].

In recent years, an open-source platform Android has been widely used in smartphones [6]. Android has a complete software package consisting of an operating system, middleware layer, and core applications. Different from other existing platforms like iOS (iPhone OS), it comes with Software Development Kit (SDK), which provides essential tools and Application

Programming Interfaces (APIs) for developers to build new applications for Android platform in Java. And also Android platform has support for Bluetooth network stack, which allows Bluetooth-enabled devices to communicate wirelessly with each other in a short distance [7]. In this paper, it aims to develop a Bluetooth-based application for the proposed home lighting control system using an open-source Android Development Tools (ADT), Android SDK (Software Development Kit) and Java Development Kit (JDK).

2. BACKGROUND AND RELATED WORK

With rapid development of information technology, the concept of Smart Living has been put forward and emerged as an attractive field for researchers and investors in the past decades. In 2006, Tom and Sitteproposed a reference model named Family System[8] which is used to describe a set of family processes, such as managing finance, planning and preparing meals, family health care, education, household maintenance, generating income and recreation and social life maintenance in Home Automation (HA), as well as their relationships, and interaction with external elements. The model of Family System can be a very useful platform for further research intocreating Smart Living to help people in daily life [8].

Bluetooth technology has been one of important technologies to home automation or Smart Living. It is a wireless technology developed to replace cables on devices like mobile phones and PCs. Although "cable-replacement" could create a point-to-point communication, Bluetooth allows wireless devices to be able to communicate with each other within range. The network of a set of Bluetooth devices is called "piconet" [9], which is anideal technology to network a smart modern home.

Recently, more and more Smart Living applications based on Android and Bluetooth have been developed [9]. Android system equips with SDK and APIs for developers to build new applications. With Bluetooth already integrated into Android system, many Smart Living systems are constructedunderAndroid system. For example, Potts and Sukittanonbuilt an Android application to lock/unlock doors remotely through Bluetooth[10]. However fora home automation system, currently many devices such as lamps and TVs don't have Bluetooth moduleembedded in the devices, so a suitable Bluetooth module and microcontrollerneed to be sought out from the marketplace [11] [12] so that a Bluetooth-based Android application can then be built using JAVA based development tool like the Eclipse or Netbeans.

3. SYSTEM DESIGN

3.1. System Architecture

In the proposed home lighting control system, a small "piconet" is established using a microchip and several Bluetooth modules BF10-A [13] [14]. The system is developed under Android platform to monitor and control home lighting via Bluetooth-enabled application.A master-slave structure is adopted in the system architecture where a Bluetooth-enabled Android phone is served as a host controller whileother Bluetooth devices, for this instance, switches, linked to the home lighting system are slave devices. The microchip controlleris set in a polling status and constantly checks any input command every 500 millisecond from the Android phone application. If it receives a command to instruct the microchip to change the lightingstatus, the microchip sends a command to the master controller through the Bluetooth module. Then the Bluetooth application executes the controls lighting operation (on or off). The detail of the system architecture is shown in Figure 1:

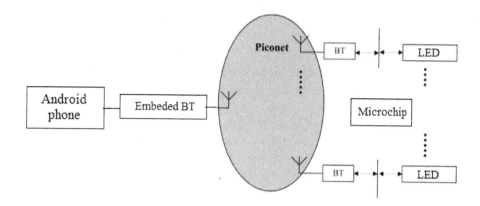

Figure 1. System Architecture

3.2. Communication Protocol

An Android phone sends its command to the client Bluetooth-enabled devices through an embedded Bluetooth module. The phone is used as a host controller which establishes their communication with Bluetooth modules via BF10-A. The communication between the master and slave Bluetooth devices covers the processes of device power-up and data exchange whereas the protocol is established in the Bluetooth software stack. The protocol layer model is specified by the Bluetooth Special Interest Group (SIG) to support the common communication between different Bluetooth devices [14].

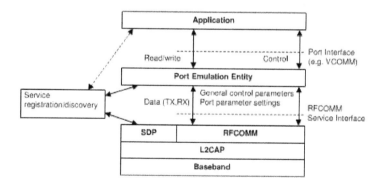

Figure 2. Bluetooth communication protocol [14]

Considering the environments and requirements of the Smart Living, the Bluetooth protocol architecture used in the application adopts the Logical Link Control and Adaptation Protocol (L2CAP), Session Description Protocol (SDP) and Radio Frequency COMMunication

(RFCOMM). Besides these protocols, an upper level protocol Serial Port Profile(SPP) is used to communicate with the application layer. Bluetooth device power-up process adopts SDP protocol to require states of the Bluetooth module while the L2CAP protocol provides data exchange service to the Bluetooth application. SSP is used on the upper level to communicate with the application layer.

3.3 Hardware Design

The overall hardware design schematic is shown in Figure 3. The slave Bluetooth module [15] BF10-A communicates with Bluetooth-enabled phone through the SPP channel, each slave module is interfaced with the lighting system with themicrochip controller which aims to decode the commands transferred from the host controller to control the lighting, and on the other side, tries to gather light status and encode it to send feedback to the master controller. UART interface P3.0 and P3.1 of the microchip is directly linked to UART_TX and UART_RX which provides the function of exchanging data. P1.0 common IO is used to control the lighting through a mediate relay in case of high current.

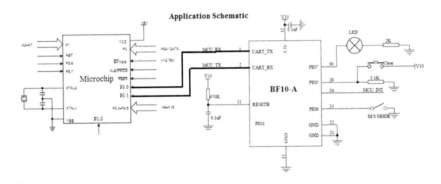

Figure 3. Hardware design schematics [13]

4. BLUETOOTH-BASED APPLICATION

The Bluetooth lighting control and monitor application provides four main functions to the phone user, namely:

- Device registration
- Lighting status monitoring
- Lighting control
- Diagnostics utility

4.1. Application Configuration

Prior to implementation of Bluetooth-based application on the phone, several software packages are required [16] [17], which include Java Development Kit (JDK), the Eclipse software environment, Android Development Tools (ADT) and Android SDK (Software Development Kit).These open-source software packages can be downloaded from:

- *www.eclipse.org* (The Eclipse)

- *developer.android.com* (ADT and SDK)
- www.oracle.com (JDK)

The Bluetooth server application is developed in Java using the Eclipse integrated development environment (IDE) which ensures the easy and quick development of the application.

4.2 IDE (Integrated Development Environment)

The proposed application on the Android phone is based on J2SE architecture and Bluetooth network technology. The entire home lighting control system consists of a master server and several clients. Followings are the steps to build the application:

- Set up Bluetooth adapter
- Find surrounding Bluetooth devices in the range
- Connect to the Bluetooth devices
- Exchange data between the master and the slave devices

The Android Bluetooth APIs are available in Java SDK *android.bluetooth* package.

4.2.1 Setting up Bluetooth adapter

The class *BluetoothAdapter* of SDK package *android.bluetooth* is used as an entry point for all Bluetooth interaction. By this, all the Bluetooth devices can be discovered. Then the Bluetooth devices are initialised according to their MAC address and finally a *BluetoothServerSocket* is created to receive echo from the surrounding Bluetooth devices. The codesused to set up a Bluetooth adapter and to enable the Bluetooth communication is listed as below [18]:

```
BluetoothAdaptermyBluetooth= BluetoothAdapter.getDefaultAdapter();
If(myBluetooth == null){Prove the found device do not support Bluetooth}
Else if(myBluetooth.isEnabled())
{Intent              enableBtIntent                  =             new
Intent(BluetoothAdapter.ACTION_REQUEST_ENABLE);
    startActivityForResult(enableBtIntent, REQUEST_ENABLE_BT);}
```

4.2.2 Finding surrounding Bluetooth modules

After setting up the Bluetooth adapter, the next step is to find the Bluetooth-enabled lighting devices by searching the matched Bluetooth modules. Before finding a device, it needs to query the list of the matched devices to make sure whether the demanded device is known to the server. The following code is used to pair devices and fetch the device name[19].

```
Set <BluetoothDevice>mypairedDevices =myBluetooth.getBondedDevices();
If (mypairedDevices.size()>0){
 For (BluetoothDevice device : pairedDevices)
 { myArrayAdapter.add (device.getName()+"\n"+ device.getAddress())} }
```

The function *getBondedDevices*returns a set of *BluetoothDevice* representing paired devices.When the query result is obtained, the codingbelow is used to find all the Bluetooth modules[20]:

```
// Create a BroadcastReceiver for ACTION_FOUND
    private final BroadcastReceivermReceiver = new BroadcastReceiver() {
```

```
    public void onReceive(Context context, Intent intent) {
        String action = intent.getAction();
        if (BluetoothDevice.ACTION_FOUND.equals(action)) {
// When discovery finds a device,
// Get the BluetoothDevice object from the Intent
            BluetoothDevice                device              =
intent.getParcelableExtra(BluetoothDevice.EXTRA_DEVICE);
            mArrayAdapter.add(device.getName()    +    "\n"    +
device.getAddress()); }}};
// Register the BroadcastReceiver
IntentFilter filter = new IntentFilter(BluetoothDevice.ACTION_FOUND);
registerReceiver(mReceiver, filter);
```

4.2.3. Establish communication between the Bluetooth devices

The Android phone must be assigned as the server in order to execute the application and the Bluetooth modules and a server-side mechanism are implemented to control lighting.Then the server opens a server socket, sends and receivescommands through the established connection. The *BluetoothSocket* class is used for the server to receive the commands when an incoming connection is accepted. The microchip used in the design plays the role of the client and it opens an RFCOMM channel to the server. The server socket listens for an incoming connection request from the clients and when one request is accepted, a *BluetoothSocket* object iscreated.

Following steps are to set up a server socket and accept a connection:

- Call *listen UsingRfcommWithServiceRecord(String, UUID)* to get a *BluetoothServerSocket* .
- Call *accept()* to Start listening to connection requests .
- Call *close*() to end the program.

4.3 Graphical User Interface design

Figure 4. Prototype GUI of the Android phone application

In order to control the home lighting, four lights are defined in the GUI. First, click the "Open Bluetooth" button to switch on Bluetooth adapter while the application is running.Click on the "Search lights" button to find the matched lights with Bluetooth devices, four devices at most, the lights automatically flash on or off according to the Bluetooth feedback received from the lights. To control the assigned light, choose either "ON" or "OFF" as shown in Figure 4, the phone then sends its command to the lights through Bluetooth communication. Finally, the user can press the "Exit" button to terminal all running threads and exit the application.

5. CONCLUSION

The objective of the paper is to realise the Smart Living, more specifically the home lighting control system using Bluetooth technology. The systemhas been successfully designed and prototyped to monitor and control the lighting status using an Android Bluetooth-enabled phone and Bluetooth modules via BF10-A. The microchip is used to assist gathering status of the lighting and provides interface to control the lighting. The Bluetooth module sends and receives commands from the Bluetooth-enabled phone and RFCOMM protocol is used in communication among Bluetooth devices. Android system JDK is used to develop the system, which is proved to be very efficient and convenient. It is concluded thatSmart Living will gradually turn into areality that consumers can control their home remotely and wirelessly[21].

REFERENCES

[1] Heidi Monson (1999)*Bluetooth Technology and Implications*, John Wiley & Sons.

[2] Cano, J.-C., Manzoni,P. and Toh, C.K. (2006). "UbiqMuseum: A Bluetooth and Java Based Context-Aware System for Ubiquitous Computing", *Wireless Personal Communications*, Vol.38, pp.187-202.

[3] Dimitrakopoulos,G., Tsagkaris,K., Stavroulaki,V., Katidiotis,A., Koutsouris,N., Demestichas, P., Merat,V. and Walter,S. (2008) "A Management Framework for Ambient Systems Operating in Wireless B3G Environments", *Springer Mobile Networks and* Applications, Vol. 13 , No. 6, pp. 555-568.

[4] Savic, S. and Shi, H. (2011) An Intelligent Object Framework for Smart Living, *Procedia Computer Science* 5 (2011) 386–393.

[5] Adams, C. and Millard, P. (2003) "Personal Trust Space and Devices: "Geography will not be History" in the m-commerce future". HawaiiInternational Conference on Business (HICB2003), 18-21 June 2003, Sheraton Waikiki Hotel, Honolulu, Hawaii, USA.

[6] Chung, C.-C., Wang, S.-C., Huang, C. Y., and Lin, C.-M.(2011) "Bluetooth-based Android Interactive Applications for Smart Living", 2011 Second International Conference on Innovations in Bio-inspired Computing and Applications *(IBICA-2011)*, Shenzhen, China, 16-18December 2011,pp.309-312.

[7] The Android open source project, http://source.android.com/.

[8] Tom, M. and Sitte, J. (2006) "Family System: A Reference Model for Developing Home Automation Applications", IEEE International Conference on Systems, Man, and Cybernetics, 8-11October 2006, Taipei, Taiwan, pp. 32-37.

[9] Piyare, R. andTazil, M. (2011) "Bluetooth based home automation system using cell phone", IEEE 15th International Symposium on Consumer Electronics (ISCE), 14-17 June 2011, Singapore, pp. 192 – 195.

[10] Potts, J. and Sukittanon, S. (2012) "Exploiting Bluetooth on Android mobile devices for home security application", Proceedings of IEEE SoutheastCon, 15-18 March 2012, Orlando, Florida, USA, pp. 1-4.

[11] Al-Ali, A.R.and AL-Rousan,M. (2004) "Java-Based Home Automation System",*IEEE Transaction on Consumer Electronics*, Vol.50, No. 2, pp. 498 - 504.

[12] Shepherd, R. (2001) "Bluetooth Wireless Technology in the Home", *Electronics & Communication Engineering*, Vol. 13, No. 5, pp.195-203

[13] BF10 Bluetooth Mould v2.0, http://www.lanwind.com/files/BF10-A-en.pdf

[14] Specifications of the Bluetooth System (Core),v1.0 B, December 1st 1999 http://grouper.ieee.org/groups/802/15/Bluetooth/core_10_b.pdf

[15] Buttery, S. and Sago,A.(2003) "Future Applications of Bluetooth", BT Technology Journal, Vol. 21, No.3, pp.48-55

[16] Sriskanthan,N., Tan,F. andKarande,A. (2002) "Bluetooth based Home Automation System", *Elsevier Journal of Microprocessors and Microsystems*, Vol. 26, pp. 281–289.

[17] Chieng D. and TingG.(2007) "High-level Capacity Performance Insights into Wireless Mesh Networking". *BT Technology Journal*, Vol. 25, Issue 2, pp.76-86.

[18] Android Tutorial | Android SDK Development & Programming, http://www.edumobile.org/android/android-development/how-to-handle-bluetooth-settings-from-your-application/

[19] Arduino, iOS, Android, and Technology Tit Bits, http://sree.cc/google/android/using-bluetooth-in-android

[20] Bluetooth Connectivity, http://developer.android.com/guide/topics/connectivity/bluetooth.html

[21] Mannings,R. and Cosier, G. (2004) "Wireless Everything - Unwiring the World", *BT Technology Journal*,Vol. 19, No. 4, pp.65-76.

ON THE SUPPORT OF MULTIMEDIA APPLICATIONS OVER WIRELESS MESH NETWORKS

Chemseddine BEMMOUSSAT[1], Fedoua DIDI[2,] Mohamed FEHAM[3]

[1,3]Dept of Telecommunication, Tlemcen University, Tlemcen, Algeria
{chemseddine.benmoussat, m_feham}@mail.univ-tlemcen.dz
[2]Dept of Computer engineering, Tlemcen University, Tlemcen, Algeria
f_didi@mail.univ-tlemcen.dz

ABSTRACT

For next generation wireless networks, supporting quality of service (QoS) in multimedia application like video, streaming and voice over IP is a necessary and critical requirement. Wireless Mesh Networking is envisioned as a solution for next networks generation and a promising technology for supporting multimedia application.

With decreasing the numbers of mesh clients, QoS will increase automatically. Several research are focused to improve QoS in Wireless Mesh networks (WMNs), they try to improve a basics algorithm, like routing protocols or one of example of canal access, but in moments it no sufficient to ensure a robust solution to transport multimedia application over WMNs.

In this paper we propose an efficient routing algorithm for multimedia transmission in the mesh network and an approach of QoS in the MAC layer for facilitated transport video over the network studied.

Keywords

Wireless mesh network, QoS, routing protocols, CBRP, CSMA/CA, 802.11e.

1. INTRODUCTION

In recent years, Wireless Mesh Networks (WMNs) attract considerable attentions due to their various potential applications, such as broadband home networking, community and neighbourhood networks, and enterprise networking. Many cities and wireless companies around the world have already deployed mesh networks. Urgently events like emergency or military forces in war for example are now using WMNs to connect their computer in field operations as well. For this application, WMNs can enable troops to know the locations and status of every soldiers or doctors, and to coordinate their activities without much direction from central command. [1]

MWNs have also been used as the last mile solution for extending the Internet connectivity for mobile nodes. For example, in the one laptop per child program, the laptops use WMNs to enable students to exchange files and get on the Internet even though they lack wired or cell phone or other physical connections in their area [2].

A wireless mesh network (WMN), as depicted in Fig. 1, consists of a number of wireless stations (nodes) that cover an area. The nodes communicate with each other in a multi-hop via the wireless links [3].

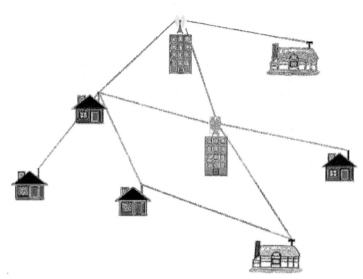

Figure 1. A generic wireless mesh network

In our work, we propose an efficient routing protocol to transport multimedia traffic in wireless mesh network and we improve MAC layer to support a real time application on WMN.

Before proposing our model, we introduce definitions of routing protocols available on the WMN and standard MAC layer to support QoS.

2. ROUTING PROTOCOLS

Generally, we can found two main types of routing protocols for wireless networks: (i) protocols which need topological information to set up a path between the nodes, (ii) protocols which require some geographical information for the route discovery process. Among these routing protocols two distinct categories can be defined:

1) Proactive like DSDV (Destination-Sequenced-Distance Vector) and OLSR (Optimized Link State Routing).

2) Reactive called also 'on-demand' like AODV (Ad-hoc On-demand Distance Vector), and DSR (Dynamic Source Routing). [4].

Short descriptions for the forth protocols listed preview are given below.

DSDV, adapted for self-configuring networks. Every node maintains its own routing table with the information about the cost of the links and network topology between the nodes.

OLSR, it uses shortest-path algorithm having the access to the routing information storing and updating periodically whenever it is needed.

AODV, it uses RREQ/RREP (Route Request/ Route Reply) mechanism for route discovery and destination SN (Sequence Numbers) for each route entry like DSDV.

DSR, it is based on RREQ/RREP packets. Like AODV protocol. However, RREQ maintains information about the whole path from the source to the destination node and gathers the addresses of the 'visited' nodes, not just the next hop. Moreover, the information is stored in a route cache instead of the routing table by every node. [4]

3. IEEE 802.11E ORIGINAL STANDARD MAC FUNCTIONS
3.1. Enhanced Distributed Channel Access and Coordination Function

The main concern of the research group, in the case of IEEE 802.11e, is to improve QoS requirements without sacrificing the interests of industry players concerned. The mechanism of distributed access, namely EDCA allows differentiation of services established at the MAC layer.
In the IEEE 802.11 (DCF), as queries are short, each occupying the network shortly, and waiting times remain still be low, the problem does not arise. However, things get complicated when transferring large files such as video or voice. To remedy these shortcomings, a new 802.11 integrating QoS, the IEEE 802.11e (EDCA), has been proposed.

The standard IEEE 802.11e aims to provide opportunities for QoS at the data link layer. It also defines the needs of different packages in terms of bandwidth and delay to allow better transmission of voice and video. IEEE 802.11e add extensions to enhance the QoS for applications with specific quality requirements, with preserving backward compatibility with variants of existing wireless networks.

The EDCA is an improvement of traditional communication mode DCF of IEEE 802.11. This protocol introduces a new concept of access category or AC (Access Category). Categories of access are: "Background", "Best Effort", video and voice. EDCA provides differentiated access and distributed to the media as well. This protocol assigns each traffic class access containing well-defined values for the parameters of DCF access. Access Media for a station depends upon the type of access associated with the stream to be transmitted. [5], [6].

The EDCA is designed for the contention-based prioritized QoS support. Table 1 show that in EDCA, each QoS-enhanced STA (QSTA) has 4 queues (ACs), to support 8 user priorities (UPs) [5] which are further mapped into four ACs

In the end, the mechanism of differentiation EDCA can provide opportunities in terms of QoS. Introduced changes at the MAC layer provide a specific treatment for each type of traffic. Research and simulations show that this differentiation ensures better transmission voice and video. However, some problems remain, such as the degradation of low priority traffic and the lack of differentiation between a call and a new call is ("handoff") [6]

TABLE 1. Access priority on different traffic in 802.11s

Priority	Access Category	Designation
1	0	Background
2	0	Background
0	0	Best Effort
3	1	Video Probing
4	2	Video at 1.5 Mbps
5	2	Video at 1.5 Mbps
6	3	Voice at 64 Kbps
7	3	Voice at 64 Kbps

3.2. Hybrid Coordination Function Controlled Channel Access

HCCA is designed for the parameterized quality of service support, which combines the advantages of DCF and PCF.

HCCA is generally considered the most advanced (and complex) coordination function. With the HCCA, QoS can be configured with great precision. QoS-enabled stations have the ability to request specific transmission parameters (data rate, jitter, etc.) which should allow advanced applications like VoIP and video streaming to work more effectively on 802.11 networks.

HCCA support is not mandatory for 802.11e APs. In fact, few (if any) APs currently available are enabled for HCCA. Implementing the HCCA on end stations uses the existing DCF mechanism for channel access (no change to DCF or EDCA operation is needed). Stations only need to be able to respond to poll messages. On the AP side, a scheduler and queuing mechanism is needed [5], [6].

4. RELATED WORK

Factors in the quality of service routing protocols become very mandatory in wireless networks because the increasing in technological advancement in these area. Getting and managing QoS in WMNs such as delay, bandwidth, paquets loss and rate error is very difficult because of the resource limitations and the complexity associated with the mobility of Mesh users and should be available and manageable

We divide our related work into two parts; the first part is summarizing the solutions into network layer and the second part we summarize a solutions and approaches in liaison layer. Finally we summarize the mains idea of each solution in a global table.

In order to provide QoS in the WMNs network the following models have been proposed:

In the beginning of the WMNs researchers started to analyze the existing routing protocols.

In [4], the principal idea is divided into two parts: first, authors compared four protocols uses in WMNs: AODV, DSR, DSDV and OLSR, with a fixed topology and other mobile, using NS -2. The results confirm that AODV protocol is the best protocol in terms of throughput, delay and that the DSR is the worst among the mentioned protocols.

Secondly, the authors introduced UDP and TCP in same scenarios of the first comparison, to assess the degree of impact of the transport layer at the network layer. The results show that UDP is more interesting than TCP in terms of quality of service management.

We can conclude that there is no ideal or best routing protocols in WMN. From the protocols studied in this paper, AODV and OLSR should be considered as the ideas worth considering. However, scalability is one of the crucial problems also in this case. One of the solutions is to propose a new routing metric for the existing protocols, use hybrid routing techniques or/and multiple radios and interfaces in order to improve performance of the network and provide better capacity of the network

With existing literature and after our previous analysis, the protocol AODV is most advantageous to ensure QoS, with this point; many works were directed towards the extension of AODV, to improve its performances. It is the aim idea of [6]. R-AODV (Rate aware routing protocol based on AODV), use minimum network layer transmission time as a performance metric. Nodes will select higher data rate link using extension of AODV.

The simulation result indicates that extension of AODV protocol can improve the throughput and decrease network delay.

For specific application like, search and rescue or emergency operations in case of natural disaster, policing and fire fighting military applications such as on the battle field, stadium, meeting rooms etc, almost all proposed routing protocols, try to converge into shortest path routing. We know that one of the advantages to use shortest path routing is that it is good for average delay in network and in overall energy efficiency because energy needed to transmit a packet is directly proportional to path length or number of hops. But a weakness of the shortest path routing is restricted to use the same nodes to route the data packets, thus causing some of the nodes to die earlier resulting into holes in the network and some of the heavily loaded nodes or even worst into partitioning of the network. Thus the need for load balanced routing emerges.

In [7] authors formulate the problem of routing as a network optimization problem, and present a general linear programming (LP) formulation for modelling the problem. Kumar and al proposes the optimized algorithm for known traffic demand and then explain the performance ratio for this. The routing algorithms derived from these formulations usually claim analytical properties such as optimal resource utilization and throughput fairness. The simulation results demonstrate that their statistical problem formulation could effectively incorporate the traffic demand uncertainty in routing optimization, and its algorithm outperforms the algorithm which only considers the static traffic demand. To achieve this objective the problem for congestion has been designed.

Overhead and bandwidth parameters are very important to have a robust network, En efficient routing protocol can solve theses problem; in next paragraphs we will summarize the recent proposed algorithm.

In [8] the global idea is to establish a route from the source to the destination that allows traffic flow within a guaranteed end-to-end latency using the minimum control overhead. Solution minimizes control overhead by effectively controlling broadcast messages in the network and it based on a reliable estimation of wireless link quality and the available bandwidth on a path routing. The quality of service awareness in the protocol is achieved by a robust estimation of the available bandwidth of the wireless channel and a proactive discovery of the routing path by an accurate estimation of the wireless link quality. Finally, the protocol uses the multi-point relay (MPR) nodes to minimize the overhead due to flooding.

In the opposite direction, from mesh nodes to Internet nodes, for all mesh nodes it exist only one direction so the gateways needs to be maintained. However, on the backward path from the Internet to mesh nodes, an individual route for every mesh node is required.

Liu et al [9] investigate protocols for this case of routing in wireless mesh networks. Using simulation experiments with realistic mobility patterns of pedestrians and cars in cities, they compare three protocols AODV, FBR and GSR, each of them represents a family of routing protocol: (i) AODV a reactive routing protocol, with an extension for mesh networks, (ii) FBR, a proactive protocol, and (iii) GSR, a source routing protocol. Their results demonstrate and confirm that an extended AODV seems to be neither scalable nor does it achieve a high packet delivery ratio; FBR has the highest packet delivery ratio but is not scalable to the network size. A good compromise is provided by GSR, which is the most scalable.

Another vision to create a solution to guarantee the bandwidth in wireless mesh network is proposed by Liu et al [10], authors proposed a QoS backup route mechanism to accommodate multimedia traffic flows in mobile WMNs and an available bandwidth estimation algorithm plus Moreover, to validate the correctness of them proposed algorithm, Liu et al implemented their algorithm on the campus wireless mesh network testbed. Their experiments and implementation show that their mechanisms can improve the network stability, throughput, and delivery ratio effectively, while decreasing the number of route failure. They implement their proposed algorithms on the testbed through an improved DSR protocol. Their implementation and experiments show that the mechanisms can effectively improve the network stability, throughput, delivery ratio, while decreasing the route invalidation ratio, and can guarantee the fluent transmission of multimedia streams.

In order to support multimedia transmission with QoS requirements, they improve the wireless routing protocol on the testbed with a dynamic ACK mechanism, which is used to balance the throughput and the quality of transmission. Additionally, authors introduce a dynamic mechanism to change the multimedia coding rate dynamically at the source node according to the available bandwidth. Moreover, they also made improvement on the admission control protocol to facilitate an experiment.

The first assertion that we can do, is that, according to the comparative studies results, done to determine what is the best choice between the existing routing algorithms in the state of the art, AODV and OLSR are the best choice by report to others, in terms of QOS.

The second assertion is that several trends have emerged, as follows:

- Extending the traditional routing algorithms such as AODV, DSR, and OLSR, to improve their performances.
- Changing values of the metric, like hybrid or dynamic metric, as bandwidth of links, or end-to-end latency instead of number of hops, for example.
- Propose protocols completely different from those present in the802.11s standard.
- Use of the clustering approach

The mesh network, as is a special case of Ad-hoc networks and MANET networks. These include a new vision of routing protocols based clusters, whose principle is very simple: divide the whole network into several parts, each party will elect a central node, responsible for coordination of routing information between other adjacent nodes, that node is named CH (Cluster Head), other nodes called its members. Communication in this type of network is simple, any member wishing to transmit, do it through its CH. The latter has a routing table, if the destination is internal (in the same group), then the delivery will be direct, if not the CH sends queries to neighbors to find the right path.

Very recent works have focused on this type of MANET routing. Mukesh Kumar [11] compared a routing protocol named CBRP (Cluster Based Routing Protocol) which gave results much interest as the basic protocols in terms of QoS (delay, throughput) and a good transition to across the MANET.

MAC protocol design is important in meeting QoS requirements since much of the latency experienced in a wireless network occurs in accessing the shared medium. In addition, MAC protocols must be interoperable with existing wireless networks operating on the same RF spectrum and fair toward all users.

Abundant hidden node collisions and correlated channel access due to multi-hop flows degrade QoS in wireless mesh networks. QoS in nearby WLANs operating on a single channel is also affected.

Mathilde Benveniste [13] propose using wider contention windows for backoff to lower the risk of repeated hidden-node collisions, a spatial extension of the TXOP concept called 'express forwarding' is an enhancement of the CSMA/CA protocol designed to reduce the latency experienced end-to-end by a multi-hop wireless mesh to clear multi-hop flows sooner, and a new mechanism called 'express retransmission' to reduce collisions on retransmission.

Simulation results show the potential benefit of the proposed enhancements and impact on fairness.

A key approach to increasing network capacity is to equip wireless routers with smart antennas. These routers, therefore, are capable of focusing their transmission on specific neighbors whilst causing little interference to other nodes. This, however, assumes there is a link scheduling algorithm that activates links in a way that maximizes network capacity. To this end, Chin et al. [14] propose a novel link activation algorithm that maximally creates a bipartite graph, which is then used to derive the link activation schedule of each router.

Authors verified the proposed algorithm on various topologies with increasing node degrees as well as node numbers. From extensive simulation studies, authors find that their algorithm outperforms existing algorithms in terms of the number of links activated per slot, super frame length, computation time, route length and end-to-end delay.

Navda et al. [15] design and evaluate Ganges, a wireless mesh network architecture that can efficiently transport real time video streams from multiple sources to a central monitoring station. Video quality suffers from deterioration in the presence of bursty network losses and due to packets missing their playback /deadline. Ganges spatially separates the paths to reduce inter-flow contention. It finds out a fair rate allocation for the different video sources.
The wireless routers in the mesh network implement several optimizations in order to reduce the end-to-end delay variation. Ganges improves the network capacity by a shortest path tree, and video picture quality by Central.

The contribution of this work [16] is twofold. First Riggio et al. propose a methodology for evaluating multimedia applications over real world WMN deployments.

Second, based on the defined methodology, they report the results of an extensive measurement campaign performed exploiting an IEEE 802.11-based WMN testbed deployed in a typical office environment. The focus of their research on three mainstream multimedia applications: VoIP, Video Conference, and Video Streaming. Two single-hop star-shaped network topologies (with symmetric and asymmetric links) and a multi-hop string topology have been exploited in order to provide a comprehensive evaluation of the testbed's performances.

For the transportation of real-time video, Moleme et al. [17] proposes a two-layer mechanism. In their solution, for channel error control ,rate adaptation is implemented in the data link layer, link stability and reliability. In addition, the network layer routing protocol is optimized for congestion control and optimal route selection by using congestion information from the data link layer and link quality metric from the network layer.

The proposed scheme aims at ameliorating the performance of UDP in WMV video streaming applications by improving throughput, packet loss and latency, so the authors in this work try to improve a standard protocol (UDP) to improve the QoS, us you know as we know, affect the operation of a standard protocol is a risk, it may have secondary effects on the proposed solutions

The framework is based on S-TDMA scheduling at the MAC layer, which is periodically executed at the network manager to adapt to changes in traffic demand. While scheduling computation is centralized, admission control is performed locally at the wireless backbone nodes, thus reducing signaling.

Leoncini et al. [18] propose two bandwidth distribution and related admission control policies, which are at opposite ends of the network utilization/spatial fairness tradeoff.

The link layer is very important to provide QoS for Wireless Mesh Networks. Researchers are focused on specific areas as we have seen. A set of researches focus on mechanisms of allocating resources such as CSMA/CA or TDMA. Other studied queue management, by doing a control admission, and another approach is to use correcting codes [19].

TABLE 2. Summarize of different approaches in WMNs

Implementations	Average delay	Over head	Packets loss	Through-put	Comments
Yinpeng Yu et al. [4]	+	+	+	+	- comparison between basic routing protocols in WMN
Zhang et al. [7]	+	-	-	+	- Improvement of AODV and comparison with the later.
Kumar et al. [8]	-	-	-	+	- Linear solution to solve short path in a critical and real applications
Sen [9]	-	+	-	+	- Establish a route that allows traffic flow within a guaranteed end-to-end latency using the minimum control overhead. - Estimation of wireless link quality and the available bandwidth are used
Baumann et al [10]	-	+	-	+	- The authors investigate protocols for backward path routing in wireless mesh networks.
Liu et al [11]	-	-	-	-	- available bandwidth estimation algorithm plus a QoS backup route mechanism - real application has been tested
Benveniste [13]	+	-	+	-	- using wider contention windows for backoff - author propose an express retransmission to reduce collision
Chin et al [14]	+	-	-	-	- An improvement of TDMA is using in place of CSMA/CA
Navda et al. [15]	-	+	+	+	- Evaluate wireless mesh network architecture that can efficiently transport real time video
Riggio et al. [16]	+	-	+	+	- Methodology for evaluating multimedia applications over real world WMN deployments.
Moleme et al [17]	+	-	-	+	- Optimization of routing protocol and mechanism of channel control
Leoncini et al [18]	-	+	-	+	- Improvement TDMA to adapt to changes in traffic demand

As we say on the beginning of our related work, we summarize all the proposed works on the following table. In this paper, the signs (+ / -) means that the authors included the chosen parameter or not among the parameters simulation in the papers.

5. OUR SYSTEM MODEL

As we say in the end of introduction, in our approaches, we propose an efficient routing protocol Q-CBRP (QoS- Clustering Based Routing Protocol) to transport multimedia traffics in wireless mesh network and we improve MAC layer to support a real time applications on WMN. We must to signal that the routing protocol is one of our approaches in [23]. The goal of this paper is to develop our proposal routing protocol with the improvement of MAC layer and to combine between the two approaches in one and only algorithm.

We will discuss in detail in this paper improvement of MAC layer to support real time applications over WMNs, and to combine between this algorithm and the routing algorithm, we create a new queues in our routing protocol, theses queues are the same in the MAC layer.

In this section, we present the basic idea of the Q-CBRP and its implementation in detail. Section 6.1 introduces the routing process CBRP briefly. In section 6.2 we define the terminology of Q-CBRP. In section 6.3 we describe and discuss about the Comparison between Q-CBRP and another's routing protocols.

After this overview we will explain our approaches in MAC layer, Section 7.1 present the improvement of MAC layer in our approaches, Section 7.2 we propose our scenario and at the last, we show results for our model.

6. OUR USES ROUTING PROTOCOL

6.1. Overview of CBRP

In generally, in sensor and MANET networks, there are several clustering protocols, among them: CBRP (Cluster Based Routing Protocol). Cluster Based Routing Protocol is an on-demand routing protocol, where the nodes are divided into several clusters. It uses clustering's structure for routing protocol.

Divides the network into interconnected substructures is clustering process that called clusters. Each cluster has a cluster head (CH) as coordinator within the substructure. Each CH acts as a temporary base station within its zone or cluster and communicates with other CHs.

CBRP is designed to be used in Wireless sensor network and mobile ad hoc network. The protocol divides the nodes of the Ad-hoc network into a number of overlapping or disjoint two-hop diameter clusters in a distributed manner. Each cluster chooses a head to retain cluster membership information. There are four possible states for the node: Isolated, Normal, Cluster-head (CH) or Gateway. Initially all nodes are in the state of Isolated. Each node maintains the Neighbor table where in the information about the other neighbors nodes is stored; CH have another table where include the information about the other neighbor cluster heads is stored. [20] The protocol efficiently minimizes the flooding traffic during route discovery and speeds up this process as well.

TABLE 3. Cluster Head Table

ID_neighbors_Clusters	ID_neighbors_Gateways	ID_members

- ID_membres : ID of all members in the same CH

TABLE 4. Gatway Table

ID_CH	ID_Members

TABLE 5. Members Table

ID_Cluster	Status	Link Status

- Status of neighboring nodes (Cluster-head, gateway or member)
- Link status (uni-directional or bi-directional)

Route discovery is done by using source routing. In the CBRP only cluster heads are flooded with route request package (RREQ). Gateway nodes receive the RREQs as well, but without broadcasting them. They forward them to the next cluster head. This strategy reduces the network traffic.

Initially, node S broadcasts a RREQ with unique ID containing the destination's address, the neighboring cluster head(s) including the gateway nodes to reach them and the cluster address list which consist the addresses of the cluster heads forming the route [21].

6.2. Terminologie for Q-CBRP

In previous works [21-22], the results show that the protocol CBRP improves QoS in mobile ad-hoc network in general. We didn't stop in this idea; so we study in detail the basic protocol to make improvements to ensure QoS in our Mesh Network.

Our improvements are summarized in two points. First we improve packet header of basic CBRP with more information to have a more complete protocol and the second point we add some fields in routing tables that we will explain in the next.

Packet ID	Source Address	Dest_ Address	List_of_visited _node	TTL	R (bps)

Figure 2. Data packet header

Figure 2 describe our proposal Data Packet Header (DPH), different to DPH in CBRP, where we add two fields in the DPH of original CBRP, the TTL (Time To Live), contains a count of number of intermediate nodes traversed to avoid the packets loop and management of the available bandwidth to guarantee QoS (R) it signifies the minimum bandwidth required by a Mesh client to transmit the data.

In our algorithm (Q-CBRP): Cluster Head Table is the same tables in CBRP protocol (Table 3) but an improvement are added in the Gateway Table (Table 4).

Gateway Table maintains the information regarding the gateway node and the available bandwidth over those nodes. We add in Gateway Table an Available Bandwidth, that mean when the data packet is sent to the destination or intermediate node it will reserve the bandwidth required by it. To perform this function of managing bandwidth, admission control mechanism is added where we also block flows when there is not enough bandwidth to avoid packets loss [23].

TABLE 6. Gateway Table in Q-CBRP

ID_CH	ID_Members	Available Bandwidth

In Q-CBRP, the Member Table maintains the information about its neighboring nodes by broadcasting a Beacon Request Packet.

6.3. QoS- Cluster Based Routing Protocol for WMN

Each node in wireless network maintains a table called Member table (Table 5) containing the address of Neighboring nodes. This table is maintained in the decreasing order of their distance from this particular node. Each node also stores the address of the Cluster-head. Cluster-head also maintains member table as well as it also maintains a gateway table which stores the address of gateway nodes in the decreasing order of distance from the centre head node. This Gateway table stores address as well as the available bandwidth of the gateway nodes.

Whenever a source node, that is member node, generates a request to transfer the data to a CH node, CH check the destination node address in it member table. If the matching node is found in the member table, packet is transferred to that node. If no match is found, then the data packet will be sent to the neighbor cluster-head. CH will again check for the match in its member table. If no match is found, cluster-head will check for the node in the Gateway node table at which the required bandwidth is available. The data packet is sent to the node at which the required bandwidth is available. The node address will be copied to List_of_Visited_Nodes field of data packet header. This field will help in the prevention of loops. Using this field, same data packet will not be sent to a particular node more than once. Reduce the available bandwidth of the gateway node. This process will continue till the destination node is reached or if the count of visited nodes get increased than the count in TTL (Time to live) field. If this count becomes more than TTL the data packet is dropped and a message is sent to source node. And finally to ensure that the packets are received in the destination and when the nodes haven't bandwidth desired by the Source, the node stop traffic for a few minutes for complete a management of the queue to avoid packet loss [23].

6.4. Discussions

The proposed protocol [23] has been implemented in the network simulator ns-2 version 2.34 [24]. The IEEE 802.11 DCF (Distributed Coordinated Function) MAC was used as the basic for the experiments with a channel capacity of 2Mb/sec.

The transmission range of each node was set to 250m. CBR is the traffic sources. The number of nodes changed with 3 values (20, 40 and 60).

In our proposed model, we chose a topology where there exist fixed nodes that represent Mesh Routers (MR) theses nodes can be CH or Gateway and mobile nodes that have a randomly circulating, theses node representing Mesh Clients MC.

Three metrics evaluated our network performances, theses metrics are: Packet Delivery Ratio (PDR), Average End to End delay (Delay) and routing Overhead (Overhead).

In [23] AODV,CBRP and Q-CBRP protocols were compared in terms of Packet delivery ratio, Average delay and routing overhead when subjected to change in pause time and varying number of Mesh clients. The results showed that by comparing the performance between Q-CBRP, CBRP and AODV, we can conclude that cluster topologies bring scalability and routing efficiency for a WMN as network size increase. By adding the management of bandwidth to our own algorithm with admission control, and add some filed in Data header plus some modification on routing Table, the mesh network is able to transport multimedia streams by offering a wider and more stable throughput compared to the basic protocol (CBRP).

7. OUR USES MAC LAYER

IEEE802.11e uses four queues with eight different priorities as mentioned previously in Table 1. For us, theses queues will not be efficient for some organizations which utilize most of their wireless networks for VoIP and video conferencing applications. According to IEEE 802.11e, two queues will be used for background and best effort data with three different priorities. Otherwise, if we consider a scenario where twenty stations are transmitting VoIP and video with one station transmitting best effort data, it will not be efficient to use two queues with three different priorities for the best effort station. In the next sections, we propose our ns-2 simulation which will overcome the mentioned limitations of the original standard when uses for VoIP and video applications.

7.1. Our Improvements

In our case, we change simulations parameters in standard IEEE 802.11e, The TOXP limit parameter is ignored in the implementation of the real network, and in our case we will demonstrate its importance

- In our approach, we used three flows (Video, VoIP and Best effort); each flow had a different data priority, we increase data priority of voice and video and we will compare with best effort data.
- We change some of the simulation parameters such as CWmin, CWmax, and AIFSN in the original IEEE802.11e standard.
- TXOP limit change varies with the priority of data.

7.2. Simulation

In our simulation, we have considered three queues to maximize the utilization of the VoIP and video applications in the network. We have also changed some of the simulation parameters such as CWmin, CWmax, and AIFSN in the original IEEE802.11e standard [24].

TABLE 7. IEEE 802.11e MAC Parameters

Parameter	Value
Slot time	20 us
Beacon interval	100 ms
Fragmentation threshold	1024 Bytes
RTS threshold	500Bytes
SIFS	20 us
PIFS	40 us
DIFS	60 us
MSDU (Voice and Video)	60 ms
MSDU (data)	200 ms
Retry limit	7
TXOP limit	3000 us

Our scenario includes a single cluster head with variable number of mobile stations moving randomly within its coverage area. The number of mobile stations is increased form 3 to 15 with three stations at a time. Every three QoS stations transmit three different types of flows (video, VoIP and best effort data) to the same destination (CH). We choose IEEE 802.11b PHY layer and Q-CBRP for routing protocol.

TABLE 8. Simulation parameters of our scenario.

Simulation parameter	Voice	Video	Best effort
Transport protocol	UDP	UDP	UDP
CWmin	3	7	15
CWmax	7	15	1023
AIFSN	2	2	3
Packet size (bytes)	160	1280	1500
Packet interval (ms)	20	10	12.5
Data rate (kbps)	64	1024	960
TXOP limit (us)	3500	3000	2500

Three metrics are evaluated in our network performances, theses metrics are: Throughput, Average Delay and ratio of packets loss.

We start with the throughput results for the first scenario, which is shown in Figures 3 and 4. In Figure3, the graph illustrates the effect of increasing the number of active QoS stations transmitting data to the access point on the throughput values for the three data flows. The sending rate in this simulation is 11 Mbps, while the CWmin and CWmax size and AIFSN values as stated in Table 8.

TABLE 9. Original IEEE 802.11e simulation parameters.

Simulation parameter	Voice	Video	Best effort
CWmin	7	10	31
CWmax	7	31	1023
AIFSN	1	2	3

In comparison, Figure 4 illustrates the effect of increasing the number of active QoS stations transmitting data to the access point on the throughput values for the three data flows using IEEE 802.11e standard [24] CW size and AIFSN values shown in Table 9.

Our CW size and AIFSN values provide better results considering the voice and video flows, but not the best effort data flow.

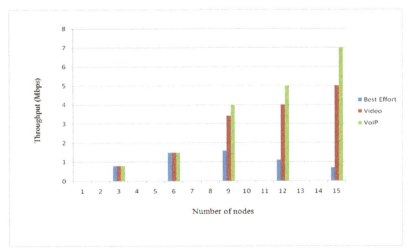

Figure 3. Simulation of Throughput using Q-CBRP with Improvement of MAC layer

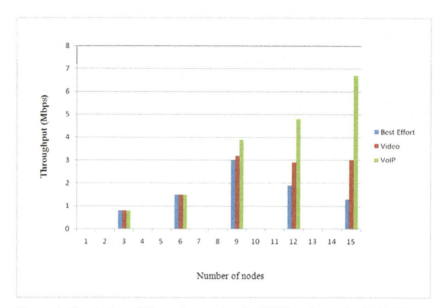

Figure 4. Simulation of Throughput using Q-CBRP with standard MAC layer

This is clearly observed from Figures 3 and 4. In both cases, it is clearly seen from the graphs that IEEE 802.11e provides service differentiation for different priorities when the system is heavily loaded by increasing the number of stations. When the number of stations is 3 or 6, all the data flows have equal channel capacity. However, in the case of 9, 12 and 15 stations, the channel is reserved for higher priority data flows. As we mentioned in the previous sections, voice flow has the highest priority among the others, while the best effort data flow has the lowest priority.

Another important factor that has a great effect on the IEEE 802.11e WLAN performance for QoS support is the packet drop and loss ratio. To calculate the number of packets dropped or lost in the transmission medium, we subtract the number of packet successfully received by the receiver (the cluster Head in our case) from the total number of packets sent by the sender (mobile stations). Table 10 shows the effect of increasing the number of active QoS stations on the packet drop and loss ratio. We vary the network load by 3 stations at a time sending three different data flows. In this simulation, we maintained the same simulation parameters in Table 8.

TABLE 10. Packet Drop ratio vs number of nodes

Number of stations	Best Effort	Video	voice
3	0 %	0 %	0 %
6	6.51 %	1.11 %	0 %
9	13.45 %	4.82 %	1.97 %
12	58.51 %	15.28 %	8.34 %
15	75.76 %	39.52 %	15.73 %

It is clearly observed from Table 10, the service differentiation between the different data flows according to their priority levels. This difference appears more when the channel is heavily loaded by increasing the number of stations. For the best effort data flow, the packet drop starts when the number of stations is 3. That is due to the fact that best effort data flow has the lowest priority. On the other hand, as the voice flow is considered, the packet drop starts when the number of stations increases to 9. This reflects the fact that voice flow has the highest priority to reserve the channel when it is heavily loaded. The percentage of the packet drop for reaches up to 76 % for the maximum channel load considering the best effort data flow, while it reaches up to 16 % for the voice flow. In fact, the system throughput is inversely proportional to the number of dropped and lost packets. In addition, packet drop has great effect on the network average end-to-end delay. Relatively, delay is directly proportional to the number of dropped packets.

The last parameter of our simulation is the average delay, delay is another important performance metric that should be taken into account. Figures 5 and 6 represent the results obtained from our simulation using different CW size and AIFSN values.

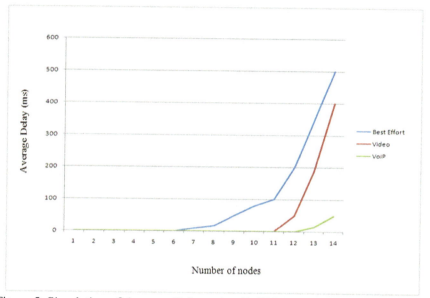

Figure 5. Simulation of Average Delay using Q-CBRP with standard MAC layer

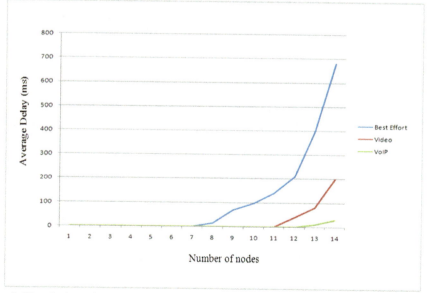

Figure 6. Simulation of Average Delay using Q-CBRP with Improvement of MAC layer

The graphs in Figures 5 and 6 illustrate the effect of increasing the number of active QoS stations transmitting data to the access point on the average end-to-end delay values for the three data flows separately from source (mobile stations) to destination CH. Our proposed CW size and AIFSN values enhances the performance with respect to the voice and video flows, but not for the best effort data flow. This is shown in Figure 6 when we have more than 12 active QoS stations. On the other hand, Figure 6 represents the simulation result using the CW size and AIFSN values in Table 8. However, as shown in this Figure 5, these values provide better results than ours with respect to best effort data flow. This is accepted for our idea, because our main concern is to enhance the performance for multimedia data flows such as voice and video.

8. CONCLUSION

We divide our paper in two proposal approach, the first approach is to use an efficient routing protocol to support multimedia application in WMNs, but for us, only efficient routing protocol in not sufficient to support a real time applications in WMNs! so we keep our routing protocol and we improve in MAC layer to had better results.

This paper compared the performance of our Algorithm Q-CBRP with improvement MAC layer in WMN and the same routing protocol with standard MAC layer. These two aspects were compared in terms of Packet loss, Average delay and Throughput.

The results show that our proposal algorithm is better in term QoS to compare with standard parameters. So we can conclude that if we combine two approaches in two level of OSI model, we have better results to compare with an approach that used only one level in OSI model.

REFERENCES

[1] XiaoHua, Xu ShaoJie, Tang Xufei, Mao Xiang-Yang Li, "Distributed Gateway Placement for Cost Minimization in Wireless Mesh Network" International Conference on Distributed Computing Systems.2010.

[2] Ian F. Akyildiz And XudongWang "Wireless Mesh Networks", Edition : WILEY , 2009.

[3] Eren G¨urses, Anna N. Kim , "Utility Optimal Real-Time Multimedia Communication in Wireless Mesh Networks" , IEEE, 2009

[4] Yinpeng Yu, Yuhuai Peng, Lei Guo, Xingwei Wang, "Performance Evaluation for Routing Protocols in Wireless Mesh Networks", International Coriference on Educational and Information Technology Performance,2010.

[5] Fedoua Didi, Houda Labiod, Guy Pujolle and Mohamed Feham "Mobility and QoS of 802.11 and 802.11e Wireless LAN standards" The International Arab Journal of Information Technology, 2009

[6] Yong Zhang, Yifei Wei, Mei Song, Junde Song "R-AODV Rate Aware Routing Protocol for Wifi Mesh Network" ICWMMN2006.

[7] Bhupendra Kumar Gupta and B.M.Acharya Manoj Kumar Mishra "Optimization of routing algorithm in Wireless Mesh Network" IEEE, 2009

[8] Jaydip Sen "A Throughput Optimizing Routing Protocol for Wireless Mesh Networks". 12th IEEE International Conference on High Performance Computing and Communications.2010p, Martin May "Routing Packets into Wireless Mesh Networks" Third IEEE International Conference on Wireless and Mobile Computing, Networking and Communications (WiMob 2007).

[9] Chungui Liu, Yantai Shu and Lianfang Zhang, Maode Ma "Backup Routing for Multimedia Transmissions over Mesh Networks" 2007 IEEE

[10] Mohammed A. Mahdi and Tat-Chee Wan "Performance Comparison of MANETs Routing Protocols for Denseand Sparse Topology" International Conference on Information and Computer Networks (ICICN 2012)

[11] Mukesh Kumar, Rahul Rishi, D.K. Madan "Comparative Analysis of CBRP, DSR, AODV Routing Protocol in MANET, IJCSE. 2010

[12] Tim Daniel Hollerung, "The Cluster-Based Routing Protocol", University of Paderborn, 2004.

[13] Mathilde Benveniste "A Distributed QoS MAC Protocol for Wireless Mesh" The Second International Conference on Sensor Technologies and Applications, 2008.

[14] Kwan-Wu Chin, Sieteng Soh, Chen Meng, " A Novel Spatial TDMA Scheduler for Concurrent Transmit Receive WMN" 24th IEEE International Conference on Advanced Information Networking and Applications, 2010.

[15] Vishnu Navda, Anand Kashyap, Samrat Ganguly and Rauf Izmailov " Real time Video Stream Aggregation in Wireless Mesh Networks" IEEE 2010.

[16] Roberto Riggio, Karina Gomez and Tinku Rasheed "On the Support of Multimedia Applications over Wireless Mesh Network" IEEE 2009.

[17] N. H. Moleme, M.a. Odhiambo, A.M. Kurien, "Enhancing Video Streaming in 802.11 Wireless Mesh Networks using

Two-Layer Mechanism Solution" IEEE 2009.

[18] Mauro Leoncini, Paolo Santi, Paolo Valente, "An STDMA-Based Framework for QoS Provisioning in Wireless Mesh Network", IEEE 2008.

[19] BEMMOUSSAT chemseddine, DIDI Fedoua, FEHAM Mohamed "A Survey on QoS in Wireless Mesh Network » MESH 2012 : The Fifth International Conference on Advances in Mesh Networks, 2012.

[20] Performance Comparison of MANETs Routing Protocols for Denseand Sparse Topology Mohammed A. Mahdi and Tat-Chee Wan International Conference on Information and Computer Networks (ICICN 2012)

[21] Mukesh Kumar, Rahul Rishi, D.K. Madan "Comparative Analysis of CBRP, DSR, AODV Routing Protocol In MANET, IJCSE. 2010

[22] Tim Daniel Hollerung, "The Cluster-Based Routing Protocol", University of Paderborn, 2004.

[23] BEMMOUSSAT chemseddine, DIDI Fedoua, FEHAM Mohamed "Efficient routing protocol to support qos in wireless mesh network", IJWMN Journal, 2012.

[24] www.isi.edu/nsnam/ns/ Network simulator ns-2, 2009.

[25] IEEE 802.11, "WG Draft Supplement to IEEE Standard 802.11-1999: Medium Access Control Enhancements for Quality of Service," IEEE802.11e/D5.0, Working Group, 2003.

CLONE-BASED MOBILE AGENT ITINERARY PLANNING USING SEPARATE TREES FOR DATA FUSION IN WSNS

Soheil Javadi[1], Mohammad H. Hajiesmaili[2], Behzad Moshiri[1] and Ahmad Khonsari[2]

[1]School of ECE, College of Engineering, University of Tehran, Tehran, Iran
s.javadi@ut.ac.ir , moshiri@ut.ac.ir
[2]School of Computer Science, IPM, Tehran, Iran
hajiesmaili@ipm.ir , ak@ipm.ir

ABSTRACT

Recent studies demonstrate that Mobile Agent (MA) approach could be more effective than conventional client-server model in Wireless Sensor Network (WSN). Particularly, itinerary planning for MAs is a significant aspect of these approaches. Tree-based methods have been widely used for this purpose where many agents are dispatched simultaneously. Most of tree-based methods try to construct an optimal/suboptimal tree in terms of an objective function. In contrast, in this paper we introduce a new approach that tries to separate the MA dispatching and data fusion operations by considering two itinerary trees which results in more flexibility to regulate the itinerary plan. Based on this idea, we propose an algorithm called Two Trees Clone-based Itinerary (TTCI) which constructs two trees, one serves to distribute MA in the network and another to fuse back the sensed data.

By experimental results, we demonstrate the performance improvement of TTCI algorithm in terms of overall energy consumption in comparison with the previous schemes. Meanwhile, the TTCI keeps the delay low.

KEYWORDS

WSN, Data Fusion, Mobile Agent (MA), clone-based itinerary planning, Energy Consumption, Delay

1. INTRODUCTION

Recent advances in wireless communication has led to rapid proliferation in development of wireless sensor network (WSN) applications [1], [2], including target tracking, battlefield monitoring, and intrusion detection [3], [4].

Each node in WSN has limited resources in terms of processing power, memory, bandwidth, and energy. In addition, a sensor node generates information that needs to be transmitted to a processing element usually called sink.

Regarding these issues, there always exists a trade-off between the accuracy of received information by the sink and amount of dispatched information [5], so we must find a way to effectively utilize precious resources. Towards this, several resource allocation schemes are proposed in the literature considering various aspects of WSNs.

To minimize the energy consumption in WSNs when employing complicated applications that required large amount of information to be transferred, a large body of research activities has been devoted to the in-network processing activities like data compression, data aggregation and data fusion as a state-of-the-art research area. In [6]-[8], several models and techniques for data fusion in the WSNs have been studied.

Data gathering in traditional DSNs usually adherences the client-server approaches, in which nodes' raw data are being sent to the processing element [9]. This approach suffers from some drawbacks: In some applications, different types of data might be produced while only homogeneous data can be fused together. In these cases, precise routing protocols are required so as homogeneous sensed data meet each other, in a right place and at a right time. Furthermore, gathering and transmission of high amount of data can result in high energy consumption and inefficient bandwidth usage.

As time went by, the concept of *Mobile Agent (MA)* arose. The idea behind the MA approach is that it might be better to send processing codes toward the nodes, instead of sending nodes' raw data to the processing element [10].

This mobile code is usually called *Mobile Agent (MA)*. MAs are software components, means code and/or execution state that can migrate in the system at run-time thus we have some kind of *code mobility*. Moreover, the program itself can make decisions about migration autonomously.

Several good reasons for using MA have been mentioned in [11], [12], such as reducing the network load, overcoming network latency, dynamic adaption. A comparison between the client-server paradigm and MA approach is presented in [13]. One of the most important reasons to use MA approach is the high flexibility in the face of requirements changing, because it is almost impossible or too difficult to reprogram sensor nodes which have already distributed in environment, while in MA approach only the processing codes require to change. On the other hand, there are still some disputes on this paradigm. One of the basic criticisms is that more resources are required to form a node, because it must be capable of executing the MA codes and the underlying framework [14].

One of the most important issues about MA approach is that each MA migrates through the network. So, the question is what is the best itinerary for MA to traverse?

This is the main focus of this work and several other research activities. Sometimes a single MA is used [5], [9], [15] and sometimes several MAs are used simultaneously [14], [16]-[19]. There is another scheme as well in the basis of cloning capability [20]. Long end-to-end delay, high energy consumption, fast energy depletion of some critical nodes, and high complexity are some of the main drawbacks that must be addressed in itinerary planning of MAs in WNS. Herein, our endeavour is devoted to propose a solution that reduces the energy consumption and keeps the delay low, while it has a special attention to reduce the fast energy depletion of the closest nodes to sink.

In this paper, we introduce a new approach towards itinerary of MAs in which two separated trees are used for dispatching MAs and data fusion, in contrast to the previous work that construct an itinerary tree for both.

The tree serves to distribute MA within network is called *Forward Tree*, and the other one employs for data fusion is called *Fusion Tree*.

Using separated trees help us to achieve more flexibility and tuning capability with regard to MA dissemination and data gathering activities.

According to the aforesaid approach, our contribution in this paper is twofold. First, we propose two heuristic algorithms for constructing the *Forward Tree*, the centralized one called *Maximum Degree Heuristic* (MDH) and its distributed version DMDH. In construction of *Forward Tree*, we try to reduce the energy consumption by minimizing the number of MA transmissions.

Second, another algorithm is proposed to create the *Fusion Tree*, which is developed to balance the load of closest nodes to the sink by building some sets and appending the unprocessed nodes to the sets with fewer members.

By putting the algorithms for constructing *Forward* and *Fusion Trees* together and in the basis of cloning capabilities of MA, we introduce an algorithm called *Two Trees Clone-based Itinerary* (TTCI).

Experimental results clearly demonstrate the performance improvement of the proposed algorithms in terms of less energy consumption and lower delay in comparison to the similar approaches.

The rest of the paper is organized as follows. Section 2 is devoted to related work. Section 3 will introduce the problem statement and proposed algorithms. Section 4 contains experimental results and finally, we conclude the paper and outline some future directions in Section 5.

2. RELATED WORK

MA-based solutions are a new trend in distributed computing, which provides more flexibility and scalability than conventional client-server models. In [21] mobile agent is represented as a middleware for automatic data fusion in WSNs and several algorithms in the context of itinerary planning have been surveyed. Although, the problem is NP-Complete [5], some previous work tried to use heuristics that yield suboptimal solutions. Note that, we only mention homogeneous WSNs (unlike [22]). Generally, one can imagine two major categories for MA itinerary planning approaches: single MA, and multiple MA approaches.

Single MA approach: In this approach, a single MA tries to traverse among all nodes and collect the data. Qi et al. in [9] proposed two heuristics called LCF and GCF. In LCF, the next node is the closest node to the current node. In GCF, the closest node to the center of environment will be the next node, so it will relocate around the center.

In LCF and GCF, only the physical distance between nodes is considered in construction of MA's itinerary plan, while one can consider many other useful criteria.

Two algorithms named IEMF and IEMA have been introduced in [15], which consider communication cost as well. IEMF operates entirely the same as LCF, but only the first node will be chosen according to the estimation of minimum communication cost. IEMA is iterative version of IEMF, where first k nodes are chosen according to the communication cost and then behaves like IEMF. It will be a trade-off between computational complexity and energy consumption.

In [5], by reducing the MA itinerary problem to a 3D traveling salesman problem, it has been shown that the problem is NP-Complete. Authors use genetic algorithm which results in a suboptimal itinerary plan.

Multiple MA approach: To address the scalability problem in single MA approach, this approach uses more than one MA, simultaneously. In [16], a genetic-based algorithm (GA-IMP) has been introduced that uses many agents simultaneously which leads to better performance than [5]. But, the problem of GA approaches mostly is related to their slow convergence rate, so they are not proper for time critical applications.

Chen et al. in [17], find the center of the dense regions (VCL), one after another. All nodes within a circle around VCL will be covered by a single MA (by methods like LCF, etc). Choosing the proper grouping parameters or dynamic radius, is not easy and highly depends on the topology. Furthermore, using a single agent in a dense area can arise the same deficiencies of single agent itinerary planning. Moreover, considering one hop communication is not a realistic assumption.

There are some approaches based on minimum spanning trees. Inspired by constrained minimum spanning tree, Gavalas et al. in [14] proposed NOID. It finds a near optimal solution where the results are satisfactory, but the computational time is high. Konstantopoulos et al. [18] improved NOID and introduced TBID, which is a greedy suboptimal algorithm. Chen et al. in [19] proposed MST-MIP. They assumed that all nodes are connected together, and weights of the edges are estimated in terms of number of hops. Then, an agent is distributed for each branch of the constructed tree. In a dense area, edges will be light weight so these edges will be selected and in this case, one agent won't be effective. Because of that, by using the hop distance to the sink, BST-MIP tries to make a trade-off between energy consumption and delay. But, if we have low density around the sink, a few agents will be dispatched and the same problem will rise again.

CBID (Mpitziopoulos et al. [20]) uses the cloning capability and tries to build a minimal tree node by node. In order to connect a new node to the nodes of the in progress tree, CBID chooses the node which leads to minimum expense. Gathering the nodes data by a single MA according to the tree, and in a depth first manner, defines the expense.

The problem is, this tree will results in some unnecessary dispatch of MAs, so increase the energy consumption. Also, it prefers deeper tree that increases the delay.

3. TWO TREES SCHEME

In this section we propose a new approach for itinerary planning of mobile agents in WSNs. As mentioned in Section 2, most of previous studies rely on building a unique tree for both dispatching the MAs and fusing the information. As a main contribution of this paper, we introduce a new approach, where we build two separate trees, one for dispatching the MAs and the second for information fusion. In this section, at first we will describe proposed algorithms to create *Forward* and *Fusion Tree*. Then by putting them together, we introduce TTCI algorithm.

3.1. Forward Tree

Transmission in a wireless channel has a broadcast-form nature, so every node in range will get the sent packet and obviously radio module will consume energy for receiving the packet which may or may not be delivered to the upper layers. Considering this property, our main goal is to distribute MA in the network with minimum number of transmissions which yields the lower energy consumption.

The nodes that are being distributed in a real topology can be divided into some levels with respect to their minimum hop distance from the sink [23]. The nodes that their minimum distance to the sink are i, construct the i th level nodes. The nodes in i th level are connected to some nodes in level i-1, i and i+1. Moreover, nodes in level i can have common neighbors in level i+1. Some of the level i nodes will send (clone) MA to some level i+1 nodes and try to cover all of them.

Our goal is to cover whole level i+1 nodes with minimum number of transmissions. Towards this, at first, we formally present this problem which we refer to as *covering problem* and then will propose some optimal and suboptimal solutions to it.

Covering problem: Suppose we have a bipartite graph consists of two sets of vertices, V (level i nodes) and U (level i+1 nodes). Each $v_k \in V$ covers (is connected to) a non-empty subset $U_k \subseteq U$. Our goal is to find a subset of V like M, such that M={ v_α | $\bigcup_{v_\alpha \in V} U_\alpha$ =U } and |M| is minimal. In the other words, there won't exists $M' \subseteq V$ that can cover U and |M|>| M' |.

In what follows, we present some algorithms for the *covering problem* to construct the *Forward tree*.

3.1.1. Naïve Algorithm

One obvious algorithm to find M, is to search through all subsets of V. It means to examine every $v_k \subseteq V$ to find out whether it covers all U or not, and finally choose the minimum one. Although this algorithm yields the optimal solution, but its complexity is exponential in the |V| ($O(2^{|V|})$). It is clear that this is too costly for big valued of |V|, so it is not a practical solution; but it can be used as a benchmark.

3.1.2. MDH Algorithm

In spite of the fact that employing the Naïve algorithm achieves us the optimal solution, it could be impractical due to its high complexity. So, we introduce *Maximum Degree Heuristic* (MDH) algorithm which leads to a near optimal solution with lower complexity.

It works as follows. In each round we choose vertex v_k among V which has the maximum neighbors among unlabeled nodes in U (unselected nodes). Then, we label unselected nodes in U that are neighbor of v_k by its index k. This process continues until all vertices in U have been labeled. In Algorithm 1 (Appendix I), you can see the procedure in more detail.

MDH is a *greedy* algorithm which will not guarantee the optimal (minimal) answers. Two examples of optimal and non-optimal solution of MDH algorithm are demonstrated in Figure 1 and Figure 2, respectively. We will see in Section 4 how effective are the obtained results from MDH, in comparison with Naïve algorithm. Due to its low complexity, we used MDH algorithm in TTCI (see Algorithm 2 in Appendix I).

Discussion on the Complexity of MDH:

At first, we assume that MDH will iterate n times to label all the nodes in U. We set v = |V| and u = |U|. Let us denote by v' and u' as average of nodes' degree in V (at the beginning) and average of nodes' degree in U, respectively. Furthermore, we suppose that each vertex in V (or U), knows its neighbors in U (or V).

The number of iterations n in the best case is $O(1)$, and in the worse case is $O(v)$. In average case, we can imagine that each vertex of V in average has v' neighbours (without any in common with other vertices). So n is equal to $O(\frac{u}{v'})$. The complexity of search and update depends on implementation. If we use a MaxHeap structure, we can find the node with largest degree in $O(1)$ and update the degrees in $O(u'v'\log v)$. Then the whole complexity (based on this consideration) is $O(\frac{u}{v'}v'u'\log v)=O(uu'\log v)=O(vv'\log v)$.

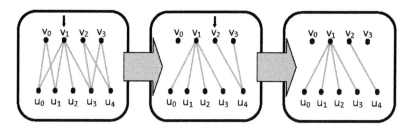

Figure 1. Optimal answer (MDH algorithm example)

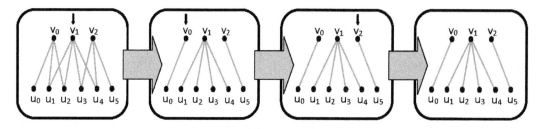

Figure 2. Optimal answer (MDH algorithm example)

3.1.3. DMDH Algorithm

In this subsection we are going to extend MDH algorithm into a distributed one named *Distributed MDH* (DMDH). At first, we assume that every node knows its level. It is not an unreasonable assumption. Consider a packet that has an additional field which is dedicated to keep the level information of the sender. The sink sends a packet which its level indicator field is initialized by zero. If a node receives a packet from a sender which is closer to the sink (in terms of hop distance) compared to the previous senders, it must update its level. Then the node must advertise its new level to its neighbors and so on. We omit further discussions about these issues because those are out of the scope of this paper. In what follows, we are going to explain the main procedure of the algorithm:

1. At the first step, every node in i th level will send a request message to its neighbor nodes.
2. In the second step, only nodes in i+1 th level introduce themselves by sending back a response message. After this phase each level i node can find out its degree (number of level i+1 neighbors).
3. Then each level i node, announces its degree. After this phase, each level i+1 node will recognize its level i neighbors' degree.
4. Finally, every level i+1 node chooses the level i node with higher degree, and announces that it has decided to connect to it. This let nodes of i level find out that to which nodes they must send the MA.

This algorithm is not optimal too but can find answers which are close to optimal. We will compare it with the optimal Naïve algorithm, in Section 4.

3.2. Fusion Tree

Construction of *Fusion Tree* can be optimized for different purposes, such as energy consumption, delay, or combination of them. There are many studies in the literature trying to build an optimal fusion (or aggregation) tree based on various criteria (even non MA-based approaches such as [24]). Note that in this work, we suppose that the construction of *Fusion Tree* is centralized and sink knows the coordination of all nodes.

Fast energy depletion of sink neighbors, especially the nodes in 1*st* level, has a significant negative impact on the network's lifetime. Thus, in our approach, our main aim is to balance the load of first level nodes, because they are the communication bridges to the sink. Fast energy depletion of some of them can lead to loss of many nodes data. One trivial solution towards this, could be to connect a new node at lower levels, to a higher level node with lower degree. But, this solution would be ineffective in some situations, because data received by a node are gathered from its entire subtree. So, a node with low degree may have large number of descendants.

Here, we are going to propose a new algorithm for building the *Fusion Tree*. Towards this, we consider each of the first level sink neighbors as a new group, i.e., the number of groups is equal to the number of direct sink neighbors. Then, at the first phase, starting from the second level, we assign each node of level i, to one of its level i-1 neighbors which its corresponding group has fewer number of members, till, all nodes are allocated to a particular group. Now, as the second phase, we try to reallocate the nodes' group to improve the overall load balancing. For this purpose, while a better allocation is found, we keep searching. Again, starting from level 2, if appending a node of level i to a node possessed in level i-1 in its neighborhood (which belongs to deferent group) leads to a lower standard deviation of groups sizes, we will rearrange the tree. Algorithm 3 (in Appendix I) shows this procedure. Obviously we can recompute the algorithm for each one of sink neighbors as a new tree root. This can balance the other nodes load also. Figure 3 is an example of the *Fusion Tree* for a sample WSN's topology, where the nodes belong to a group are demonstrated in the same color.

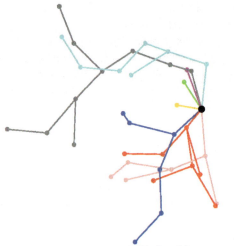

Figure 3. Optimal answer (MDH algorithm example)

3.3. TTCI Algorithm

So far, we proposed some algorithms for building *Forward* and *Fusion Trees*. In this subsection, we are going to introduce the *Two Trees Clone based Itinerary* (TTCI) algorithm by putting them together. This approach can provide us more flexibility and possibility of adjustment between different design criteria. Generally speaking, the structure used in the most tree based itinerary planning approaches is as follows:

1. Building *Fusion Tree*
2. Distributing MAs according to the constructed tree
3. Fusing the information based on the constructed tree

On the other hand, our approach contains one more phase:

1. Building *Forward Tree*
2. Building *Fusion Tree*
3. Distribution of MA according to *Forward Tree*
4. Fusing the information back to the sink, according to *Fusion Tree*

The algorithm works as follows: All nodes within the sink transmission range will constitute the first level nodes. For the sake of simplicity, we assume the sink and ordinary nodes have the same transmission range.

Then, unvisited nodes that are directly reachable from the level i nodes, will constitute the level i+1 nodes. This will continue until we visit all nodes and construct all level sets. Fortunately, this can be done concurrently with building the trees, so, there is no need for extra processing. We used MDH algorithm to create *Forward Tree* level by level (Algorithm 2 in Appendix I). Also, we have introduced Algorithm 3 (in Appendix I) for building *Fusion Tree*. After the phases of tree construction, MA can start its itinerary. Each MA cloner node, clones itself (the MA), and waits for arrival of data produced by its children through *Fusion Tree*. These cloned MAs will be annihilated after delivering data to their parent.

Obviously MA needs to carry the information about its itinerary. Simply, each node has to know, to whom it should send and from who it should receive. We use two trees instead of one, so, we must send extra information compare to single tree approach. Employing the hierarchical scheme, allows us to prune the itinerary information effectively. When a level i cloner node receives the MA, it can prune the itinerary information corresponding to nodes with level$\leq$i-1. Thus, in an ordinary tree based approach, every node has to send the itinerary information of size N (number of the sensor nodes). But in our approach, each node (v_i) has to send the itinerary information of size $\sum_{j=v_i\text{-}level}^{j \leq \max Depth} N_{l_j}$ in which N_{l_j} is the number of j th level sensor nodes.

In WSNs, communication is not completely reliable. Also, nodes may face some problems which will not let SN to operate properly. It means that when the delay of receiving data from a certain node exceeds a particular threshold, we should proceed without that. Here, we do not mention the issues about this timer settings and robustness. We suppose that we are not facing the failure and mobility of the nodes. Many researchers consider limited amount of energy for nodes, to study the network's lifetime. Also, some works consider sensor malfunction which leads to elimination of sensors, to study the robustness of their algorithms in deal of sensor death. But, these are not our concerns in this study. We didn't consider the collision condition as well.

4. EXPERIMENTAL RESULTS

In the following section, we will present experimental results to show the effectiveness of the algorithms proposed in this paper. All implementation codes were written in Java. Moreover, results have been reported by 95% confidence level. At first, we will show the experimental results about algorithms introduced to build *Forward* and *Fusion Tree*. Then we benchmark TTCI with another clone-based algorithm called CBID [20].

Energy model used in the simulation is based on the fact that the energy consumed by radio module has 3 states: Tx, Rx, and sleep [25]. Also, there are another models in which energy consumption is proportional to the number of bits [24], [26], so we reported the number of sent/received bytes. Some have used a simple linear model (in terms of the number of bits) for processing and receiving [26]. Also, sending energy can be considered as it is proportional to the square of the distance. It means $E_S \propto d^2$ [24].

Table 1. Simulation parameters

Parameter	Value
Area	500×500 (m^2)
Transmission range	45 (m)
Transmission rate	250 (kbps)
MA code size	1 (KByte)
Local data	2 (KByte)
Fusion factor (output/input)	0.7
Tx energy consumption	57.42 (mW)
Rx energy consumption	62 (mW)
Periodic energy consumption	6 (mW)

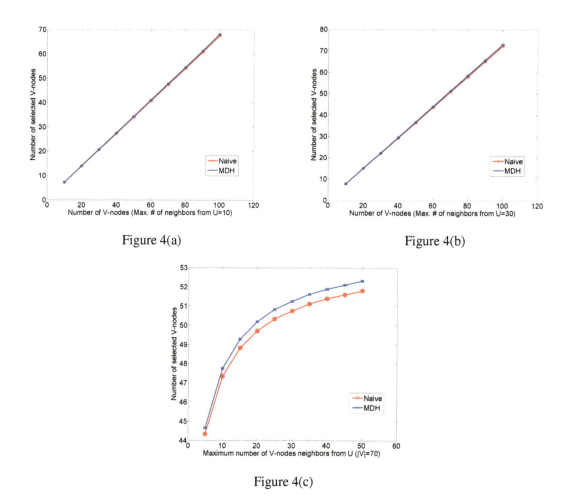

Figure 4(a)

Figure 4(b)

Figure 4(c)

Figure 4. Naïve and MDH comparison

4.1. Comparison of Naïve, MDH, and DMDH

To compare Naïve and MDH, we examined the results over different topologies. Figures 4(a) and 4(b) show the effect of increasing the number of level i nodes, in the situation of fix average number of neighbors in level i+1. On the other hand, in Figure 4(c) the number of level i nodes has been fixed, while average number of neighbors (in level i+1) is variable. Results from Figure 4(a), Figure 4(b), and Figure 4(c) clearly demonstrate that results of two algorithms are similar and very close. Therefore, due to lower complexity, employing MDH instead of optimal Naïve can make sense. Thus, in TTCI, we used MDH to build *Forward Tree*.

You can see the phrase "*Maximum number of neighbours*" in some figures, which needs further explanation. We choose a number Max≤|U| to build a random topology which means that each node in V will have at most Max neighbours in U. Then a random number $Rand_{v_k}$ will be chosen for node v_k in V such that $Rand_{v_k} \leq Max$ is equal to the number of v_k's neighbours in U. For example, in Figure 4(c), Max=50 shows that each node in level i has 25 neighbours in average (because the distribution is uniform). It is obvious that by increasing the Max, the topology becomes denser. Also, we compare the results of Naïve and DMDH in Figure 5. We can see that the results are close and can lead to proper selection of V-nodes for the *covering problem*.

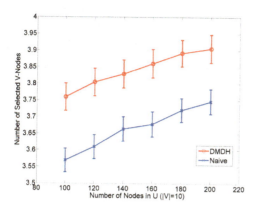 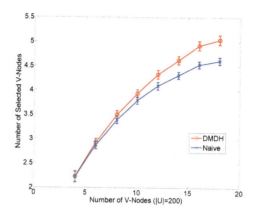

Figure 5(a). Effect of increasing the U-nodes   Figure 5(b). Effect of increasing the V-nodes

Figure 5. Naïve and DMDH comparison

4.2. On the Performance of Fusion Tree Algorithm

Figure 6(a) shows the effectiveness of our *Fusion Tree* algorithm on balancing the number of nodes attached to each group. You can see that even after first phase in our algorithm, we have a good balance compared to the tree constructed by CBID. Analyzing the improvement phase of our algorithm is not straightforward because of the randomness in the topology. Thus, for the sake of better illustration of the complexity, we report the mean number of changes in Figure 6(b), which means the maximum number of the *while* loop (line 17 of Algorithm 3 in Appendix I) iteration.

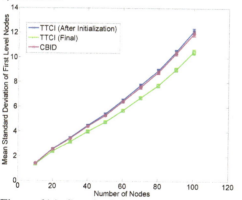 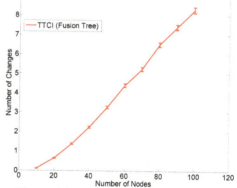

Figure 6(a). Standard deviation of groupCount   Figure 6(b). Improvement loop of TTCI: Number of changes

Figure 6. Performance of fusion tree algorithm

4.3. Simulation and Implementations outputs

Now we present the result of simulation that compares TTCI with CBID [20]. We used Castalia 3.1 framework [25] (built upon OMNeT++ simulator [27]). Table 1 summarizes the simulation conditions. Also, we mentioned the overhead of itinerary plan carried by MA in both algorithms. Delay reported for 3 rounds of data gathering.

As it was expected, TTCI has less delay, because CBID may has deeper subtrees (Figure 7(a)). Figure 7(b) shows the average of energy consumption. It shows the superiority of proposed algorithm. Some packets fail because the radio module of receiver is not in Rx mode. Figure 8(a) shows the aforesaid packet breakdown. The difference between received packet in TTCI and CBID shows that, CBID tends to send more packets compare to TTCI. Finally, Figure 8(b) shows the average number of bytes sent/received in each algorithm. TTCI has superiority because it uses the minimum number of cloners. Moreover, CBID cannot prune its itinerary data, so it has to send more packets.

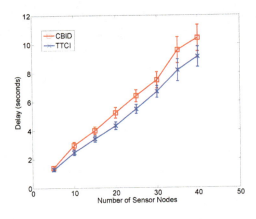

Figure 7(a). Delay comparison

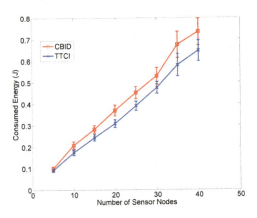

Figure 7(b). Mean energy consumption comparison

Figure 7. Comparison of TTCI and CBID

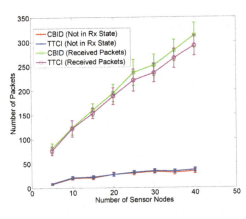

Figure 8(a). Received packets

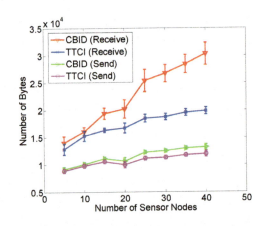

Figure 8(b). Packet exchange comparison

Figure 8. Comparison of TTCI and CBID

5. CONCLUSION

Due to importance of in-network data processing such as data fusion and data aggregation, several studies have focused on its efficiency. MA approaches can reduce energy consumption. They also increase flexibility in the face of requirements changing.

Lots of researches have been conducted to find the optimal itinerary plan for MA. In this paper, we introduced a tree-based approach according to the cloning capability of MAs, which uses separate trees for distributing the MAs and performing data fusion. Based on this scheme, a new algorithm named TTCI is presented. Different algorithms were studied in order to build *Forward* and *Fusion Trees*. Simulation results revealed the effectiveness of our approach in reduction of energy consumption and keeping the delay low. There are several future directions to extend this work.

One future work will be studying this method in environment along with collision conditions. Robustness is another aspect that could be addressed in the future work. Finally, different algorithms can be used to build the trees for further improvement based on different criteria.

REFERENCES

[1] I.F. Akyildiz, W. Su, Y. Sankarasubramaniam, and E. Cayirci, "A Survey on Sensor Networks," *IEEE Communications Magazine*, vol. 40, pp. 102-114, 2002.

[2] N. Xu, "A survey of Sensor Network Applications," *IEEE Communications Magazine*, vol.40, pp. 102-114, 2002.

[3] P. Santi, *Topology Control in Wireless Ad Hoc and Sensor Networks*, John Wiley & Sons Ltd, 2005.

[4] J. Yick, B. Mukherjee, and D. Ghosal, "Wireless Sensor Network Survey," *Computer Networks*, vol. 52, pp. 2292-2330, 2008.

[5] Q. Wu, N.S.V. Rao, J. Barhen, S.S. Iyengar, V.K. Vaishnavi, H. Qi, and K. Chakrabarty, "On Computing Mobile Agent Routes for Data Fusion in Distributed Sensor Networks," *IEEE Trans. Knowledge and Data Engineering*, vol. 16, no. 6, pp. 740-753, Jun. 2004.

[6] E.F. Nakamura, A.A.F. Loureiro, and A.C. Frery, "Information Fusion for Wireless Sensor Networks: Methods, Models, and Classifications," *ACM Computing Surveys (CSUR)*, vol. 39, no. 3, 2007.

[7] S.K. Das, *High-Level Data Fusion*, Artech House Publishers, 2008.

[8] H.B. Mitchell, *Multi-Sensor Data Fusion: An Introduction*, Springer, 2007.

[9] H. Qi, and F. Wang, "Optimal itinerary analysis for mobile agents in ad hoc wireless sensor networks," *Proc. Int'l Conf. on Wireless Communications*, pp.147-153, 2001.

[10] H. Qi, S.S. Iyengar, and K. Chakrabarty, "Multi-Resolution Data Integration Using Mobile Agents in Distributed Sensor Networks," *IEEE Trans. Systems, Man, and Cybernetics, Part C: Applications and Reviews*, vol. 31, no. 3, pp. 383-391, 2001.

[11] D.B. Lange, and M. Oshima, "Seven Good Reasons for Mobile Agents," *Communications of the ACM*, vol. 42, no. 3, pp. 88-89, 1999.

[12] C.G. Harrison, D.M. Chess, and A. Kershenbaum, "Mobile Agents: Are they a good idea?," *Mobile Object Systems Towards the Programmable Internet*, vol. 1222, pp. 25-45, 1997.

[13] Y. Xu, H. Qi, and P.T. Kuruganti, "Distributed Computing Paradigms for Collaborative Processing in Sensor Networks," *IEEE Global Telecommunications Conference, GLOBECOM '03*, vol. 6, pp. 3531 - 3535, Dec. 2003.

[14] D. Gavalas, A. Mpitziopoulos, G. Pantziou, and C. Konstantopoulos, "An approach for near-optimal distributed data fusion in wireless sensor networks," *Wireless Networks*, vol. 16, no. 5, pp. 1407-1425, 2010.

[15] M. Chen, V. Leung, S. Mao, T. Kwon, and M. Li, "Energy-Efficient Itinerary Planning for Mobile Agents in Wireless Sensor Networks," *IEEE Int'l Conf. Communications, 2009. ICC'09.*, 2009.

[16] W. Cai, M. Chen, T. Hara, L. Shu, and T. Kwon, "A Genetic Algorithm Approach to Multi-Agent Itinerary Planning in Wireless Sensor Networks," *Mobile Networks and Applications*, pp. 1-12, DOI:10.1007/s11036-010-0269-z, 2010.

[17] M. Chen, S. Gonzlez, Y. Zhang, and V.C.M. Leung, "Multi-Agent Itinerary Planning for Sensor Networks," *Proceedings of the IEEE 2009 International Conference on Heterogeneous Networking for Quality, Reliability, Security and Robustness (QShine 2009)*, Las Palmas de Gran Canaria, Spain, 2009.

[18] C. Konstantopoulos, A. Mpitziopoulos, D. Gavalas, G. Pantziou, "Effective Determination of Mobile Agent Itineraries for Data Aggregation on Sensor Networks," *IEEE Trans. on Knowledge and Data Engineering*, vol. 22, no. 12, pp. 1041-4347, Dec. 2009.

[19] M. Chen, W. Cai, S. Gonzalez, and V.C.M. Leung, "Balanced Itinerary Planning for Multiple Mobile Agents in Wireless Sensor Networks," *Ad Hoc Networks*, vol. 49, no. 7, pp. 416-428, 2010.

[20] A. Mpitziopoulos, D. Gavalas, C. Konstantopoulos, and G. Pantziou, "CBID: A Scalable Method for Distributed Data Aggregation in WSNs," *International Journal of Distributed Sensor Networks*, DOI:10.1155/2010/206517, Hindawi Publishing Corporation, 2010.

[21] A. Mpitziopoulos, D. Gavalas, C. Konstantopoulos, and G. Pantziou, "Mobile Agent Middleware for Autonomic Data Fusion in Wireless Sensor Networks," In M.K. Denko, L.T. Yang, and Y. Zhang, *Autonomic Computing and Networking, chapter 3 (pp. 57-81)*, USA:Springer, 2009.

[22] G. Wu, H. Li, and L. Yao, "A Group-based Mobile Agent Routing protocol for multitype Wireless Sensor Networks," *IEEE/ACM International Conference on Green Computing and Communications & IEEE/ACM International Conference on Cyber, Physical and Social*, pp. 42-49, 2010.

[23] S. Kulkarni, A. Iyer, and C. Rosenberg, "An Address-Light, Integrated MAC and Routing Protocol for Wireless Sensor Networks," *IEEE/ACM Transactions on Networking*, vol. 14, pp. 793-806, 2006.

[24] H.O. Tan, I. Korpeoglu, and I. Stojmenovic, "Computing Localized Power-Efficient Data Aggregation Trees for Sensor Networks," *IEEE Trans. Parallel and Distributed Systems*, vol. 22, no. 3, pp. 489-500, Mar. 2011.

[25] A. Boulis. (2010). Castalia A Simulator for Wireless Sensor Networks and Body Area Networks. version 3.1 User's Manual, NICTA. [online]. Available: castalia.npc.nicta.com.au/pdfs/Castalia\%20-\%20User\%20Manual.pdf.

[26] S. Eswaran, M. Johnson, A. Misra, and T. La Porta, "Adaptive In-Network Processing for Bandwidth and Energy Constrained Mission-Oriented Multi-hop Wireless Networks," *In Proceedings of the 5th IEEE International Conference on Distributed Computing in Sensor Systems*, pp. 87-102, DOI:10.1007/978-3-642-02085-8_7, 2009.

[27] A. Varga. (2004). OMNeT++ Discrete Event Simulation System. version 3.0 User Manual. [online]. Available: www.omnetpp.org/doc/omnetpp/Manual.pdf.

APPENDIX I

The Appendix I contains the pseudo code of the presented algorithms. These may further clarify the algorithms' purpose.

Algorithm1: MDH Algorithm
Definitions
1: V and U are two array of nodes (in the bipartite graph) which we want to cover whole U by some of the nodes in V
2: int[] **deg** keeps unprocessed degree of V-nodes (deg[k]=unprocessed degree of v_k which is initialized by its degree)
3: **index** is an integer for tracing the next selected node in V
4: M is the resulting suboptimal set
5: findMax(deg) returns the index of element with the maximum value in deg
6: N(v_k) returns all neighbors of v_k
7: $u_k \cdot parent$ is the node in V which is selected to cover u_k
Main

```
1:  while ∃u_k ∈ U such that u_k · parent is null do
2:      index ← findMax(deg)
3:      M · add(u_index)
4:      for each u_k' ∈ N(v_index) such that u_k' · parent is null do
5:          u_k' · parent ← v_index
6:          for each v_k" ∈ V such that u_k' ∈ N(v_k") do
7:              deg[k"]--
8:          end for
9:      end for
10: end while
```

Algorithm 1. MDH Algorithm for the Covering Problem

Algorithm 2: Forward Tree Algorithm

Definitions
1: V is the array of all nodes of size n+1 (do not getting confused by V and U in Alg. 1)
2: int[] **deg** for unprocessed degree of each node initialized by 0 (deg[k]=unprocessed degree of v_k, k=1, ···,n)
3: v_0 is the sink
4: **processed, candidates** and **children** are set data structures (also, they can be implemented by ordinary arrays)
5: v_{index} is used to point to a selected node (temporary variable)
6: N(v_k) returns the neighbors of v_k
7: parent of a node, is the node which is selected to cover it
8: findMax(deg,candidates) returns the node with maximum degree in candidates set
9: v_k · addChild() adds a node to the children list of v_k

Main
```
1:  processed ← v_0
2:  candidates ← N(v_0)
3:  for each v_j ∈ candidates do
4:      v_j · parent ← v_0
5:      v_0 · addChild(v_j)
6:  end for
7:  while processed · size()<(n+1) do
8:      for each v_j' ∈ candidates do
9:          for each v_j" ∈ N(v_j') do
10:             if v_j" ∉ (candidates ∪ processed) then
11:                 children ← children ∪ { v_j" }
12:                 deg[ j' ]++
13:             end if
```

```
14:         end for
15:       end for
16:       if children · size()>0 then
17:         while ∃$v_{k'}$ ∈ children such that $v_{k'}$ · parent is null do
18:           $v_{index}$ ← findMax(deg,candidates)
19:           for each $v_{k''}$ ∈ N($v_{index}$) such that (($v_{k''}$ ∈ children) &&
                                                          ($v_{k''}$ · parent is null )) do
20:             $v_{k''}$ · parent ← $v_{index}$
21:             $v_{index}$ · addChild($v_{k''}$)
22:             for each $v_\alpha$ ∈ N($v_{k''}$) such that $v_\alpha$ ∈ candidates do
23:               deg[$\alpha$]--
24:             end for
25:           end for
26:         end while
27:         processed ← processed ∪ candidates
28:         candidates ← children
29:         children · clear()
30:       end if
31: end while
```

Algorithm 2. Forward Tree Algorithm used in TTCI (based on MDH)

Algorithm3: Fusion Tree Algorithm

Definitions
1: **maxDepth** shows the maximum depth of the *Forward Tree*
2: V is the set of all nodes as in Alg. 2
3: sink (v_0) level is 0 and we assume that by execution of Alg. 2, each node level is determined
4: **FuParent** of a node (v_k), is the node which is selected to receive its (v_k) data for data fusion (and the same for addFuChild())
5: v_k and $v_{k'}$ are temporary variables
6: **groupCount** is an array of integers of size |N(v_0)|
7: group is the groupID assigned to a node, initialized by -1
8: N(v_k) returns all neighbors of v_k
9: FuRemove($v_{k''}$) removes $v_{k''}$ from the fusion children list, of a node

Main
1: for each node like v_k ∈ N(v_0) do
2:     assign a new groupID (from [0 .. |N(v_0)|-1]) to v_k · *group*
3:     v_k · *FuParent* ← v_0
4:     v_0 · addFuChild(v_k)
5:     groupCount[v_k · *group*]++
6: end for
7: for i from 2 to maxDepth do

8:     **for** each node like v_k which $v_k \cdot level==i$ **do**
9:         find the best $v_{k'}$ which $(v_{k'} \in N(v_k))$ && $(v_{k'} \cdot level==i-1)$ && (groupCount[$v_{k'} \cdot group$] is minimum)
10:        $v_k \cdot FuParent \leftarrow v_{k'}$
11:        $v_{k'} \cdot addFuChild(v_k)$
12:        $v_k \cdot group \leftarrow v_{k'} \cdot group$
13:        groupCount[$v_k \cdot group$]++
14:    **end for**
15: **end for**
16: boolean **ctnu** $\leftarrow$ true
17: **while** ctnu==true **do**
18:    ctnu $\leftarrow$ false
19:    **for** i from 2 to maxDepth **do**
20:        **for** each node like v_k which $v_k \cdot level==i$ **do**
21:            find the best $v_{k'}$ which $(v_{k'} \in N(v_k))$ && $(v_{k'} \cdot level==i-1)$ && (appending v_k to $v_{k'}$ leads to minimum standard deviation of groupCount)
22:            **if** $v_{k'} \neq v_k \cdot FuParent$ **do**
23:                $v_k \cdot FuParent \cdot FuRemove(v_k)$
24:                $v_k \cdot FuParent \leftarrow v_{k'}$
25:                $v_{k'} \cdot addFuChild(v_k)$
26:                ctnu $\leftarrow$ true
27:                decrease groupCount[$v_k \cdot group$] by the number of v_k descendants +1
28:                increase groupCount[$v_{k'} \cdot group$] by the number of v_k descendants +1
29:                update $v_k \cdot group$ and its descendants with $v_{k'} \cdot group$
30:            **end if**
31:        **end for**
32:    **end for**
33: **end while**

Algorithm 3. Fusion Tree Creation Algorithm used in TTCI

PERFORMANCE EVALUATION OF MAC PROTOCOLS FOR AD-HOC NETWORKS USING DIRECTIONAL ANTENNA

Arvind Kumar[1], Rajeev Tripathi[2]

Department of Electronics and Communication Engineering, MNNIT, Allahabad
arvindk@mnnit.ac.in[1], rt@mnnit.ac.in[2]

ABSTRACT

An Ad-hoc network is a dynamic network farmed on demand by a group of nodes without any pre-existing network infrastructure. Self configurability and easy deployment feature of the Mobile Ad-hoc network (MANET) resulted in numerous applications in modern era. An efficient and effective medium access control (MAC) protocol is essential in ad-hoc network for proper sharing of channel. The omni-directional antenna has been used in traditional MAC protocols. More intensive attention is given toward the directional antenna by researchers due to capability of spatial reuse and other beneficial features. Main focuses of this paper is to discus and evaluate the performance of MAC protocol using omni-directional and directional antenna. We conducted a comparative simulation study of different MAC protocols using Omni-directional and directional antenna. Performance metrics like throughput and delay are used for the performance analysis. On the basis of result derived from simulation a comparison among these MAC protocols using directional antenna is given.

Keywords

Ad-hoc network, Directional antenna, Directional MAC protocol, Smart antenna.

1. INTRODUCTION

There are two possibilities for enabling wireless communications: infrastructure mode and ad hoc mode. The first one relies on infrastructure that needs to be built in advance. In 802.11 infrastructure mode, all the wireless devices in the network can communicate with each other through an Access Point (AP) or communicate with a wired network as long as the AP is connected to a wired network. The other choice is ad hoc mode whose major feature is the non-existence of supporting infrastructure. An ad-hoc network is self organized wireless network without fixed or backbone infrastructure. Ad-hoc network topology is changes dynamically due to mobile nature of node hence routing protocols are required for data communication between source and destination. All nodes have routing capability and use peer-to-peer packet transmission or forward packet for other node using multi-hop communication. Due to flexibility in deployment, ad-hoc networks are very useful in military and other application such as emergency and rescue operation where infrastructure is unavailable or unreliable. Wireless mesh networks could be considered as a type of wireless ad hoc networks. Mesh networks extend the reach of wireless networks and are ideally suited for many environments such as commercial zones, neighbourhood communities and university campuses. Wireless sensor networks are another application of ad hoc networks in which sensor devices are connected in open peer-to-peer ad hoc network architecture to offer various utilizations such as monitoring traffic congestion in a city, detecting a biological weapon in the battle field and border intrusion.

Traditional work on ad hoc networks assumes that each device is equipped with omni-directional antennas. With the continuing reductions in the size and cost of directional antenna in recent years, it has become feasible to use directional antennas for ad hoc networks. Directional antenna offer potential benefits for wireless ad-hoc networks. With directional transmission and reception, spatial reuse ratio and antenna gain can be increased substantially; this leads to significant improvement in communication system performance. To best utilize the directional antenna, a suitable MAC protocol is required. Directional antenna based MAC protocols are capable of transmitting only in certain narrow azimuth that significantly reduces the chance of collision and increase the effective network capacity. However, use of directional antenna introduce the some complex issues like hidden terminal, exposed terminal and deafness problem at the same time trans-receiver complexity increases.

Carrier sense medium access (CSMA) [1] protocol causes hidden terminal and exposed terminal problems. These problems are minimized by MACA (medium access collision avoidance) [2] using RTS/CTS (request to send/clear to send) frames. Some optional controls are added for better performance in Multiple Access with Collision Avoidance for Wireless (MACAW)[3]. IEEE 802.11 MAC DCF protocol [4] is CSMA/CA (Carrier sense multiple access with collision avoidance) with optional RTS/CTS control message. The IEEE 802.11 MAC protocol is designed to exploit omni-directional antennas and could not work well in directional antenna based ad hoc networks. Therefore, several modified MAC protocols have been proposed to exploit directional antennas, enhance the spatial reuse, and increase network capacity.

This paper is organized as follows: In Section II brief description of related works already done in area of MAC protocol using directional antenna is given. Section III describes the basic concepts, operation and type of smart antenna. Architecture and operation of MAC protocol is given in Sections IV. Performance evaluation is given in Section V that discusses the simulation setup and simulation result under the static and dynamic topology. Finally, we conclude with future scope of works in Section V.

2. RELATED WORKS

In this Section we review some directional MAC protocols proposed in literature for ad-hoc networks using directional antenna. The major advantage of directional antenna with 802.11 based ad-hoc network is the reduced interference and the possibility of having parallel transmission to increase the spatial reuse of radio resources [5]. A number of directional protocols have been proposed for wireless ad-hoc networks using directional antenna.

IEEE 802.11 DCF MAC [6] is originally designed for omni-direction transmission using omni-directional antenna known as CSMA/CA. If mobile node senses the channel is free/ idle it waits for duration DIFS, if channel is still idle it will wait for additional random duration of time called back off time. Once back off time expires, data transmission is initiated. After each collision CW (Contention window) size is doubled up to maximum window size. On the other hand after each successful transmission minimum size of window is set. If data packet size is larger than specific threshold, it will use four ways hand shake mechanism for data transmission. If mobile node has packets to transmit, it will transmit the RTS frame (RTS frame having the identification of intended receiver), in reply of RTS receiver will transmit CTS frame. After receiving the CTS frame source will send the DATA packet, after successful reception of DATA, receiver will send the ACK frame for each data packets. These control packet (RTS, CTS) decrease the probability of data collision, as they allow communicating nodes to reserve the channel for the entire communication duration before the actual data transmission begins. All neighbours nodes of sender or receiver are expected to keep silent to avoid collision or interference with ongoing transmission, which causes low spatial reuse. The basic operations of these schemes are illustrated in figures 1(a) and (b).

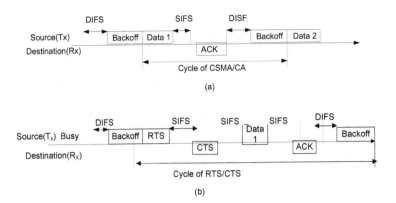

Figure 1. Timing diagram of (a) CSMA/CA and (b) RTS/CTS

In [7] authors proposed a MAC protocol using directional antenna. It has been assumed that each node knows it own position and neighbour's node, and directional and onmi-directional transmission range are same. Two schemes: directional RTS (DRTS) MAC and omni-directional RTS (ORTS) MAC have been proposed. Node will send the ORTS frame onmi-directionally if none of the directional antenna elements are blocked, otherwise node will send a DRTS frame, provided that desired directional antenna element is not blocked. In both scheme the DATA/ACK are transmitted directionally, while the CTS is transmitted omni-directionally, if it does not interfere with other ongoing transmission. Nodes have different transmission range for directional and omni-directional transmission in MAC protocol proposed in [8]. All frames are transmitted directionally. RTS frame send using multi-hop while CTS, DATA and ACK are transmitted using single hop. In [9] authors proposed an algorithm in which RTS, CTS, DATA and ACK frames are transmitted using the directional antenna in the directions which are free according to the directional network allocation vector (D-NAV) table. [10] Presents a MAC protocol using omni-directional RTS/CTS and directional DATA, to reduce the level of interference. In this scheme mobile node do not have any location information.

In [11] authors have proposed dual busy tone multiple accesses (DBTMA) and its directional version (DBTMA/DA) in [12]. It divides the channel in two sub channel. Busy tone is transmitted omni-directionally in [11] while directionally in [12]. Authors have split the channel in two sub channels [13], one is used for RTS/CTS/DATA/ACK and the other is used for busy tone transmission. A Selective Directional MAC (SDMAC) protocol is proposed by authors in [14]. This paper deal with the problems, like deafness problem, hidden terminal problem, and Heal on line (HOL) blocking problem. A scheduling algorithm has been proposed to deal with HOL blocking problem. A range adaptive directional MAC is proposed in [15]. In the place of single fold directional transmission range they proposed multi fold transmission range. A range based DNAV and distance between source and destination is used to select the transmission range. In [16] authors give the comparative study of different omni-directional and directional MAC protocols like 802.11 MAC, MACA, MACAW, FAMA, DBTMA and DBTMA/DA.

3. SMART ANTENNA

The improvement of the spectral efficiency in Ad-hoc network may be achieved through the application of directional antennas. The employment of directional antennas has demonstrated improvement in the performance of Ad-hoc network. This enhancement is because directional antennas concentrate the power into limited regions, and consequently, they considerably diminish the interference caused to users that are not within these regions. However, directional antennas lack of the required flexibility that ad-hoc networks demand. Automatically re-configuration of the network is violated by the employment of directional antennas. A more appropriate kind of antennas that have the advantages of directional antennas and offer a major flexibility is the so-called smart antennas. Smart antennas generally combine multiple antenna elements with a signal processing capability to optimize its radiation and/or reception pattern automatically in response to the signal environment. Basically, there are two major categories of smart antennas.

3.1. Type of Smart antenna

3.1.1. Switched beam antenna

Switched beam antenna is comprised of multiple fixed beams that are formed by shifting the phase of each antenna element of an antenna array by a predetermined amount, or simply by switching between several fixed directional antennas. The transceiver can select one or more beams to transmit or receive.

3.1.2. Adaptive arrays antenna

Adaptive arrays antenna is theoretically able to form an infinite number of radiation patterns. The patterns are created taking into account the desired signal and the interferers. In other words, they have the capability of direct the main beam toward the desired signal while suppressing the antenna pattern in the direction of the interferers. Adaptive arrays tend to perform better than switched beam antennas, since they place the desired signal at the maximum of the main lobe and reject the interferers. Nevertheless, adaptive arrays are not convenient for ad-hoc networks due to their high complexity and cost. Switched beam antennas, although not performing at the same level of adaptive arrays, offer the advantages of directional antennas joint to a major flexibility, and a lower cost and complexity compared to adaptive arrays.

3.2. Need of Smart Antenna

There are many motivations to utilize the smart antenna technique in a wireless system. As an example of a cellular communication system where the capacity has become a critical issue, the use of conventional omni-directional antennas not only causes huge waste of signal energy because only a small part is transmitted to desired receiver but also generates serious interference to neighbouring base stations and terminals. Therefore, dividing one cell into several sectors and uses a directional antenna for transmission was developed with the goal of reducing the interference level. Dividing in sectors has shown ability to increase frequency spectrum utilization. However, sectorized systems lack the ability to change the antenna's beam-width or orientation in response to a changing propagation environment and traffic condition. This shortcoming results in large capacity waste in sparse traffic sectors and traffic blocks in dense traffic sectors. The adaptive array smart antenna system which can intelligently control its radiation pattern based on signal processing provides an excellent solution to these problems. Its feature of generating null towards interferers results in higher frequency spectrum utilization and thus increases the system capacity.

The smart antenna technique is applicable for almost all major wireless protocols and industrial standards to achieve larger network coverage and higher system capacity. Examples of these standards could be FDMA (frequency division multiple access) employed in AMPS(advanced mobile phone system) and NMT (Nordic mobile Telephone); TDMA(time division multiple access) employed in GSM (global system for mobile communication) and IS-136; CDMA (code division multiple access) employed in IS-95, WCDMA (wideband CDMA) and TD-SCDMA(time division synchronous CDMA); FDD(frequency division duplexing) and TDD (time division duplexing). As its costs continue to decline, the smart antenna offers a practical, economical solution to address wireless network capacity and performance challenges for different communication systems, including RFID, WiMax, Ultra wideband (UWB), and even WiFi.

3.3. Antenna model used for simulation

In this paper we used the switched beam antenna. Each node have N highly directional, fixed predefined antenna beam elements which are deployed into non overlapping fixed sectors, each spanning an angle of 360/N degree. Each node can transmit in two modes: omni-directional and directional mode. That has been presented in figure.2 by unicast and broadcast transmission.

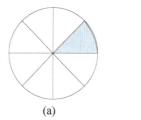
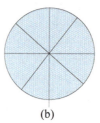

(a) (b)

Figure 2. Azimuth pattern of 8 sector directional antenna for (a) unicast and (b) multicast

When a node n forms a transmission beam at an angle θ and a beam width Φ with a transmission range R, the coverage range of node n at an angle θ is defined as transmission_zone$_n$ (θ, Φ, R) as shown in Fig. 3.

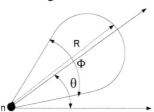

Figure 3. Transmission Zone of a node n (transmission_zone$_n$ (θ, Φ, R)), with beam width Φ, beam angle θ and transmission range R

Without loss of generality, we assume that the direction of each beam is fixed and the boresights of the first sector are always directed towards the $0°$ and Φ on a polar plane, in this case θ will be $0°$. When a link needs to be established for communication between nodes i and j, then node i calculates the relative angle, ϕ_{ij}, between the $0°$ in the polar plane and link (i, j) to determine the employed antenna beam. Based on the above assumptions, the selected beam ζ_{ij} that node i would use to communicate with node j via link (i, j) is given by equation given below.

$$\zeta_{ij} = \left\lceil \frac{\phi_{ij}}{\Phi} \right\rceil$$

Node j will apply the exact same procedure to determine the beam ϕ_{ji}. Figure 4 shows the distribution of the antenna beams and the beam selected by node i to establish link (i, j).

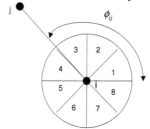

Figure 4. The selection of the beam that is used for link (i, j) is based on angle ϕ_{ij} which depict which beam will be used for this communication link.

4. MAC Protocol

Despite recent advances in MANET, there is still a wide range of different open research issues. One of the most important building blocks of wireless ad-hoc networks consists in designing an efficient medium access control (MAC) scheme. A well designed MAC protocol is essential to maximize the performance and the efficiency of the network. One approach for multiple accesses is to employ contention based schemes where nodes compete for accessing the channel. However, contention-based medium access method are inherently inappropriate for providing QoS guarantees, which is becoming a basic premise for lots of communication. Collision-free access techniques are another kind of MAC scheme more suitable for QoS wireless ad-hoc networks, since it is possible to guarantee QoS. All above discussed MAC protocols are based on using omni-directional antenna.

Directional antenna has been broadly used in various communication systems. However, to simply use directional antenna with the conventional IEEE 802.11 standard for ad hoc network could not bring substantial network improvements. This encourages many researchers to develop new MAC protocols which could fully exploit the advantages of directional antennas.

4.1. MAC layer problems using directional transmission

The protocols mentioned in introduction sections allow the spatial reuse and performance enhancement. However, these methods fails to address the issues arise using the directional transmission, like deafness problem, hidden terminal problem and HOL blocking problem.

4.1.1. Deafness problem

Deafness occurs when a node attempts to transmit to a node, which is already busy in transmission in another direction [7]. Deafness is due to mainly directional RTS/CTS messages. The destination node using a directional antenna is deaf in all directions except for the direction of transmission/reception; such a node does not receive, and hence, does not respond to any RTS requests from other directions. In most cases, the attempting node is also deaf to the destination's transmission/reception, since it cannot sense the ongoing directional communication. In the literature, some solutions are proposed to alleviate the deafness problem with directional transmissions. Sub-band busy tone [8] and circular MAC [10] prevent from defenses problem in Ad-hoc network using directional antenna.

4.1.2. Hidden terminal problem

A hidden terminal can be defined as a terminal that is not aware of the ongoing communication between transmitter/receiver pairs and whose intended transmission could lead to the failure of the ongoing transmitter/receiver pair's communication. Conventional MAC protocols for ad hoc networks (the IEEE 802.11 operated in ad hoc mode using -directional antenna) address the hidden terminal problem by employing the RTS/CTS handshaking mechanism before data transmission. Using directional antenna at physical layer, RTS/CTS packets are transmitted directionally. Using directional antenna hidden terminal problem is due to unheard RTS/CTS. Hidden terminal problem results from the feature of a directional antenna that its antenna gain towards a desired direction is larger than the gain towards other directions. Multiple access and busy tone solution are presented in [1].

4.1.3. HOL (Head of Line Blocking) problem

This effect occurs because it is possible for the medium to be free in some directions but not others. Head of Line blocking phenomenon is common in First in First Out (FIFO) nature queues. This problem becomes significant in wireless ad hoc networks using directional antennas as it uses FIFO queue which consists of packets intended for different directions. A packet on the top of the queue may block the remaining packets if it finds the medium busy in its intended direction, where as the packets in the queue intended for other directions may find the medium to be idle. This problem reduces the performance of the protocol to a great extent.

4.2. Architecture

This section details the architecture and operation of MAC protocol. In this architecture every node keeps a MAC table that gives the blocked and unblocked beams and best beam for transmission between source and destination. All nodes send and receive the packet directionally. We assume every node know the relative direction of its neighbouring nodes. This information can achieved through GPS system installed at each node. Switching between beams and blocking of the beams are done via radio frequency switches. At the receiving node selective diversity is applied to determine the best beam over which the highest SNR is measured. Beam information is receded and updated during the RTS and CTS frame exchange, frame format are depicted in figure 4.

RTS/CTS frame of given MAC protocol has one and three extra fields than the frame defined in IEEE802.11 MAC protocol. "Beam Number" is used to indicate the beam number used by the transmitter to send RTS/CTS packets. "Best beam number" indicate the best beam on the basis of SNR to be used by the transmitter and receiver. Other fields have the same meaning as in IEEE802.11 MAC frame. This concept had been given in [20] and used for wireless mesh networks but with four antenna beams. We have used this concept with eight antenna beams to improve the MAC performance when number of node is large which is presented as Angular-MAC with eight beams "A-MACEB".

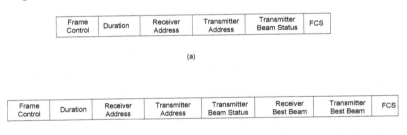

Figure 4. (a) RTS and (b) CTS frame format

4.3. MAC Operation

When node A has data for transmission to node B, if it senses channel is free, then it will transmit RTS frame using all beams. RTS [B, A, n] packet are sent over all beam and they are received over the beams of surrounding nodes as shown in figure 5. Where n is used to represent the 8 beam starting from 1 to 8 and 8 bits are being used and bit position value in bit pattern represent the each beam. Bit pattern 00000100 represent that this packet is being transmitted using beam number 3 (3rd bit is 1). RST [B, A, 8] indicate that RST packet is transmitted to node B using beam 8 of node A and it is being received by the node B over beam 5. Other RTS transmitted by the beams of A, may be received by the node B, but on the basis of SNR value beam 5 is selected. Node C and D also receive the RTS packet transmitted because they are in transmission range of A but they will not reply. C and D record the beam number used by A as best beam for communication to A that can be used in future. After receiving the RTS intended to B, it will sent CTS packet using all beams. CTS [A,B,4,8,4] indicate that B is source and A is destination of CTS packet , B use the beam number 4, beam number 8 and 4 are best beam of node A and B respectively for transmission between A and B. CTS [A, B, n, 8, 4] are sent by all beam of node B to its surrounding. Node C also receives the CTS [A, B, 4, 8, 4] over its beam number 7. Node C will block its beam 4 and 7 for transmission between A and B because

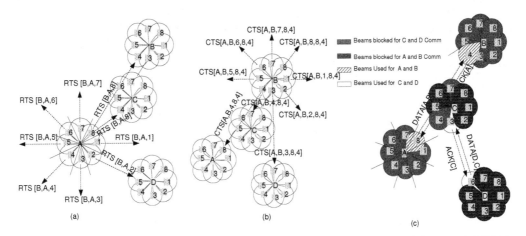

Figure 5. Beam utilization and signal transmitted for RTS (a) and CTS (b) and DATA and ACK(c)

these beams are on the transmission path of A and B. Node A and B will block their beams except best beam. After receiving CTS from node B, A and B will exchange the DAT and ACK using the beam 8 and 4 of node A and B respectively. Meanwhile, if node D want to communicate to node C, they will exchange the RTS and CTS using the unblocked beams and select the best beam for transmission. After the RTS and CTS exchange, node C will block the remaining beams except beam number 3 and similarly node D will block all the beams except beam 6 that are selected as best beam for transmission between C and D as depicted in figure. Each node maintains a table that has information of all nodes and their blocked and unblocked beams, best beams at that node.

5. PERFORMANCE EVALUATION

5.1. Simulation Setup

Simulation is done on Qualnet [21] that is written using PARSEC [22], a C language base discrete event simulator. Traffic model used in our simulation is CBR with packet size of 512 byte and transmission data rate of 2 Mbps. The simulations are run for different random seed and the results are statistically averaged out for 25 iterations, each running for 600 simulation seconds. For fair comparison of MAC protocols we have used two different topologies, first is 5X5 mesh topology with different source and destination combination.

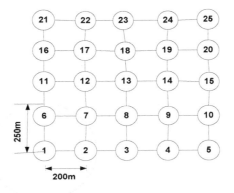

Figure 6. 5X5 mesh network scenario used in simulation

Row and column are being separated by 200 meters, as given in figure.6 and table.1. Transmission range of each node is 250 meters. Second topology is random, in which 25 nodes are randomly placed and simulation is done for different source and destination pair for 600 simulation seconds. We have used the AODV (Ad hoc on demand distance vector) routing protocol. Performance of IEEE802.11 MAC, D-MAC, Angular-MAC with eight beams "A-MACEB" are compared with performance matrices like throughput.

Table1. Simulation Parameters

Parameters	Values
Topology	Mesh, Random
Number of node (mesh , random)	5X5, 25
Row spacing (mesh)	200 meters
Column spacing (mesh)	200 meters
Channel frequency	2.4 GHz
Transmission range	250 meters
Data rate	2 Mbps
Receiving threshold	-81.0 dBi
CS threshold	-91.0 dBi
Main lobe antenna gain	12.0 dBi
Number of beams	8
Packet size	512 bytes
CBR packet arrival rate	2 ms to 600ms
Simulation time	600 sec.

5.2. Simulation Results

In first Scenario we consider the single hop communication between any two node e.g. connection between nodes 1 to 2 (1 → 2) and connection between nodes 6 to 11 (6 →11). The performance of proposed MAC (Angular-MAC with eight beams "A-MACEB") is higher compared to IEEE802.11 MAC but nearly same as D-MAC and A-MACFB. This is due to fact that, with IEEE802.11 two connections cannot transmits the packets at the same time but using the D-MAC, A-MACEB simultaneous transmission is possible due to multiple beam antenna of each node. Performances of A-MACEB increased with node density, because multiple beam antennas are under utilized in the case of low density but they are efficiently utilized with high node density at the cast of increased processing time. The second scenario used, two multi-hop communication between any node e.g. connection between nodes 1 to 21 (1 →21) and connection between nodes 1 to 5 (1→5). Performance of D-MAC and A-MACEB are better compared to IEEE802.11, but performance of these MAC is nearly equal.

Figure 7 shows the aggregated throughput in third scenario where 10 simultaneous (1→5, 6→10, 11→15, 16→20, 21→25, 1→21, 2→22, 3→23, 4→24, and 5→25) communications are present in network. This illustrates throughput as function of CBR data sending rate for mesh topology. The A-MACEB achieves approximately 40% improvement in throughput as compared to the 802.11 MAC and 10% improvement compare to DMAC. In random topology improvement is more compare to mesh topology due to random location of node during simulation this is given in figure 8. Figure 9 and figure 10 show the performance of A-MACEB , DMAC and 802.11MAC for throughput against delay. A-MACEB that gives the performance improvement in both cases but it is better in the random topology.

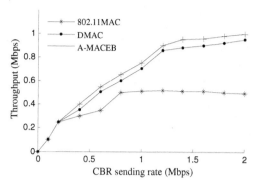

Figure 7. Aggregated throughput comparison of different MAC in mesh topology

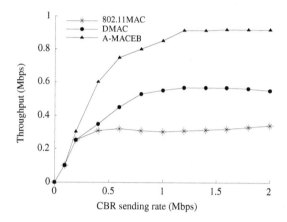

Figure 8. Aggregated throughput comparison of different MAC in random topology

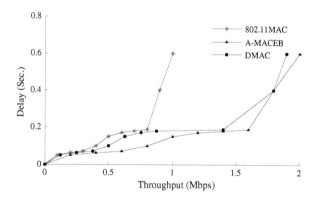

Figure 9. Delay comparison for different throughput in mesh topology

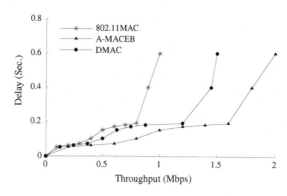

Figure 10. Delay comparison for different throughput in random topology

6. CONCLUSION

We have performed comparative analysis of omni-directional and directional MAC for ad-hoc networks. We studied the basic concepts of smart antenna and use of directional antenna in ad-hoc networks. Further we discuss the MAC protocol in directional antenna and issues like deafness, hidden terminal and HOL problems. We have done simulations to evaluate the performance for CBR traffic on mesh and random topology. The results illustrate that directional MAC protocol A-MACEB and DMAC are more effective than omni-directional MAC 802.11MAC, especially when traffic load is high. It is noticed that A-MACEB achieves the better performance compare to the DMAC.

REFERENCES

[1] F. A. Tobagi and L. Kleinrock, (1975) "Packet switching in radio channels part II the hidden terminal problem in carrier sense multiple access and busy tone solution", *IEEE Transaction on Communication*, pp 1417-33.

[2] Phil Karn,(1990) "MACA – A new channel access method for packet radio", in *ARRL/CRRL amateur radio 9th computer networking conference*, pp 134-140.

[3] V. Bharghavan, Alan Demers, Scott Shenker and Lixia Zhang, (1994)"MACAW: A media access protocol for wireless LAN's", in *ACM SIGCOMM*, London, UK, pp 212-125.

[4] "Wireless LAN medium access control (MAC) and physical layer (PHY) specifications", 1997, *Draft standard IEEE802.11, P802.11/D1*: The editors of IEEE802.11.

[5] G. Li, L.L. Yang, W.S. Conner, B. Sadeghi, (2005) "opportunities and challenges for Mesh Networks Using Directional Antennas", *WiMESH'05* , Santa Clara, Califorina, USA.

[6] "Wireless LAN medium access control (MAC) and physical layer (phy) specifications", *IEEE standard working group*, 1999.

[7] Y.B. Kuo, N.H. Vaidhya, " Medium access Control protocols using directional antenna in adhoc networks", *Proceeding of IEEE INFOCOM*, Tel-Aviv, Israel. March 2000.

[8] R.R. Choudhary, R. Ramanathan, N.H. Vaidhya, (2002)"Using directional antenna for medium access control in adhoc networks", *Proceeding of ACM MobiCom*, Atlanta, Georgia, USA, pp 23-28.

[9] K. Kobayashi, M. Nakagaw,(2000) " Spatially Divided channel scheme using sectored Antenna for CSMA/CA Directional CSMA/CA", *Proceeding of IEEE PIMRC*, London.

[10] T. Korakis G. Jakallari and L. Tassiulas, (2003)" A Mac protocol for full exploitation of directional antenna in ad-hoc networks", *Proceeding of the 4th ACM International symposium on mobile Ad hoc networking and Computing (MobiHoc)*, pp 98-107.

[11] J. Den and Z. Haas,(1998) "Dual busy tone multiple access (DBTMA): A new medium access control for packet radio networks", in *ICUPC*, Florence, Italy.

[12] Zhuochuan Huang, Chien-Chung Shen, Chavalit Srisathapornphat, and Chaiporn Jaijaeo,(2002) " A busy-tone based directional MAC protocol for ad-hoc networks," in *IEEE MILCOM*, Anaheim, CA.

[13] R. R. Choudhury, N. H. Vaidhya, (2004)"Deafness: A mac problem in ad-hoc networks when using directional antennas," *Proceeding of the 12th IEEE international Conference on networks Protocols (ICNP)*, pp. 283-192

[14] Pan Li, Honhqiang Zhai, Yuguang Fang, (2009)" SDMA: Selective directional MAC protocol for wireless mobile Ad-hoc networks," *ACM Transactions on networking*, Vol. 7, pp. 1106-1117.

[15] Kai Chen, Fan Jiang, (2007)"A range-adaptive directional MAC protocol for wireless ad hoc networks with smart antennas," *International Journal of Electronics and Communications*, Volume 61, Issue 10, Pages 645-656.

[16] Z. Huang and C.-C. Shen,(2002) "A Comparison Study of Omni-directional and Directional MAC Protocols for Ad Hoc Networks," *Proc. IEEE Globecom*.

[17] R. Janaswamy, " Radio wave propagation and smart antenna for wireless communications," Kluwer Acadmic publisher.

[18] T. Rappaport ,(2002) *Wireless communications principles and practice*, Prentice Hall.

[19] R. Ramanathan, (2001) " On the performance of ad-hoc networks with beam forming antenna", In *proceeding of the 2nd ACM international Symposium on Mobile Ad hoc Networking and Computing (MobiHoc)* , pp 95-105.

[20] E. Ulukan, O. Gurbuz ,(2008) "Angular MAC: a frame work for directional antennas in wireless mesh networks", *Journal on Wireless Networks*, Vol. 14 Issue 2, pp 259-275.

[21] Qualnet Simulator Version 5.0, Scalable Network Technologies,www.scalable-networks.com.

[22] UCLA parallel Computing Laboratory, PARSEC, http://pcl.cs.ucla.edu/, 1999.

On Using Multi Agent Systems in Cognitive Radio Networks: A Survey

Emna Trigui, Moez Esseghir and Leila Merghem_Boulahia

Autonomic Networking Environment, ICD/ERA, CNRS UMR STMR 6279
University of Technology of Troyes, 12, rue Marie Curie, 10010 Troyes Cedex, France
`{emna.trigui, moez.esseghir, leila.merghem_boulahia}@utt.fr`

Abstract

In the last decade, cognitive radio technology received a lot of consideration for spectrum optimization. This issue creates huge opportunities for interesting research and development in a wide range of applications. This paper presents a state of the art on cognitive radio researches especially works using multi-agent systems. We propose among others a classification of cognitive radio proposals based on multi-agent concept and point out the pros and cons for each of the described approach.

Keywords

Cognitive radio, multi-agent systems, radio spectrum management, wireless network

1. Introduction

Wireless networks have shown an impressive progression these past few years. Their fast evolution arise a much larger need for spectrum resources. Consequently, new efficient approaches for spectrum resources allocation must be implemented. Cognitive radio (CR) technology has been introduced as a key concept of dynamic spectrum resources allocation. CR was conceived to operate across different spectrum bands and heterogeneous radio access technologies (RAT). To achieve cognitive radio goal, which consists in improving spectrum allocation, recent works investigate different methods and protocols such us game theory, medium access control (MAC) protocols and multi-agent systems (MAS). As each approach is related to a specific protocol layer (MAC, network or application layers) and handles a particular CR management function (sensing, sharing, etc), we end up with a large variety of solutions [1].

Using multi-agent systems (MAS) seems to be one of the approaches that suits well to the cognitive radio specifications. First, it guarantees the autonomy of users as embedded agents can manage their own spectrum need in a dynamic and decentralized manner. Second, agents can perceive their environment and communicate with each other, which is mandatory for a CR terminal. Third, the intelligent property of an agent leads to smart MAS and so to an efficient Cognitive Radio Network (CRN).

Various multi-agent system based approaches were proposed to sense the spectrum holes on the one hand, and to allow information sharing and decision distribution among multiple CR terminals, on the other hand. The underlying key idea of using MAS in CRN is to manage fairly and in a decentralized way the shared radio resources between multiple cognitive radio users, in order to enhance the overall spectrum efficiency.

2. COGNITIVE RADIO OVERVIEW

Cognitive radio concept has been recently introduced in order to enhance the efficiency of the radio spectrum usage in next generations of wireless and mobile computing systems. Basically, the cognitive radio offers new opportunities for resolving static spectrum sharing problem by allowing CR nodes to sense unused spectrum bands and dynamically access them. In the following subsection, we provide a brief introduction to cognitive radio concept. Then, we focus on CR functions and we explain how relevant they are to spectrum management.

2.1. Cognitive Radio Concept

Cognitive radio concept was firstly introduced in 1998 by Joseph Mitola III [2] as *"a radio that employs model-based reasoning to achieve a specified level of competence in radio-related domains"*. A cognitive radio changes its transmission or reception parameters in such a way to avoid interferences between users and to enhance the entire wireless communication network's efficiency. A CR terminal interacts with its radio environment, senses and detects free spectrum bands and then uses them opportunistically. Accordingly, it has enough capabilities to effectively manage radio resources.

In cognitive radio networks, there are two types of users: licensed or primary users (PUs), and unlicensed or secondary users (SUs). PUs can access the wireless network resources according to their license while SUs are equipped with cognitive radio capabilities to opportunistically access the spectrum. Cognitive radio capability allows SUs to temporarily access the PUs' under-utilized licensed channels. To improve spectrum usage efficiency, cognitive radio must combine with intelligent management methods. In the following subsection, we first describe CR primary functions and then, we detail proposed solutions to improve dynamic spectrum access.

2.2. Spectrum Management

Cognitive radio system requires four major functions [3] that enable it to opportunistically use the spectrum. These functions consist in the CR terminal's main steps for spectrum management. They are: spectrum sensing, spectrum decision, spectrum sharing, and spectrum mobility. Each function is defined as follows.
1) Spectrum Sensing: It is the key enabling technology for cognitive radio networks. Its main objective is to detect the unused channels and to provide more spectrum access opportunities to CR nodes without interfering with PUs.
2) Spectrum Decision: This function is needed to select the best channel detected through the sensing phase, according to specific cognitive radio user's requirements (such as usage time, quality of service, width of spectrum band).
3) Spectrum Sharing: It is the allocation of the selected spectrum band amongst coexisting primary and secondary users while avoiding interferences between them.
4) Spectrum Handoff or Mobility: When an agreement with PU for spectrum sharing comes to end, CR terminal has to switch towards another band. It performs a spectrum handoff.

Throughout this paper, the expressions CR node (or CR terminal) and secondary user (or SU) will be used interchangeably for the rest of the document. Likely, the words access, allocation and sharing will also be used interchangeably.

Figure 1 illustrates the four main spectrum management functions of the cognitive radio cycle as well as the possible transitions between them. After performing spectrum sensing, CR node has to choose the most appropriate channel among detected free ones according to its application's requirements: it is spectrum decision. Next, CR terminal starts spectrum sharing process. Here, two transitions are possible: going back to sensing at the end of a sharing agreement or switching immediately to another band (spectrum mobility) in case a PU begins to use the same current channel for example. Spectrum mobility can happen proactively or reactively. In the first case, CR node predicts periodically the target band. In the other one, it initiates sensing when handoff is needed.

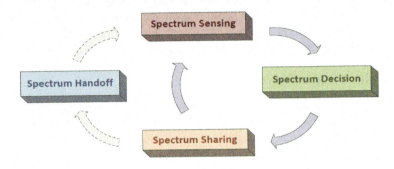

Figure 1. Spectrum management functionalities

Different approaches are proposed in the literature to provide effective solutions for each spectrum management function described above.

For example, regarding the spectrum sensing issues, some studies have proposed radio frequency energy detection [4, 5] which goal is to detect the presence of PU based on the signals that CR node can detect. Few other studies have investigated the hypothesis that empty spectrum portions are a combination of PU's signals, the additive white Gaussian noise and the signal gain [6, 7]. Other researchers have drawn their focus towards matched filter detection [8, 9]. In this technique, PU's signals are previously known and the corresponding match filter generates a high value of gain, which maximizes the received signal-to-noise ratio (SNR).

Another category of studies have focused on game theory [10] to achieve efficient spectrum sensing and spectrum sharing tasks, respectively. In those approaches [11-14], SUs form coalitions and sense cooperatively the spectrum in order to identify and access fairly free channels. Each user has a payoff calculated according to its participation in the coalition's tasks.

Medium Access Control (MAC) based solutions have been recently developed for dynamic spectrum access [15, 16] to improve overall spectrum efficiency. MAC protocols act on spectrum sensing, sharing and mobility. Some of them, referred as cognitive sensing, exploit sensing stimuli to build up a map of the spectrum opportunities that CR terminals can use. Other MAC protocols are concerned with spectrum sharing. They can help to schedule available resources, and distribute them upon CR users. The main goal of other MAC protocols is spectrum mobility. They aim to allow cognitive users to vacate selected channels when their quality becomes unacceptable.

MAS based approaches [36-57] are also increasingly used to insure dynamic spectrum access in CRN. MAS solutions have been proposed mostly in order to address the issues of spectrum sensing and spectrum sharing. These proposals will be discussed in section IV.

Spectrum handoff is relatively a new area of research and only a few investigation efforts have been done in the recent past. In those studies, two main spectrum handoff schemes were proposed: reactive and proactive. Through reactive approaches [17, 18], SUs perform spectrum switching after detecting the arrival of a PU in the band, the target channel is then selected instantaneously. However, through proactive spectrum handoff approaches [19-23], SUs predict the channel availability status and perform spectrum switching before a PU occupies the channel. This prediction is based on previous channel usage statistics. As an example, in [22] is proposed a predictive model for dynamic spectrum access based on the past channel history.

Compared to the reactive spectrum handoff, the proactive approach may be able to reduce handoff delay because the channel is preselected. Nevertheless, it can face a big challenge in case where the preselected target channel is no longer available when the spectrum handoff procedure is started.

In this paper, we will focus on using MAS for the spectrum management in CRN since the application of this technique in such a domain is increasingly occurring. Researchers use MAS in cognitive radio context as it offers a distributed, interactive and intelligent system added to the large range of autonomous and intelligent mechanisms that are already provided. Moreover, the high similarity between an agent in MAS and a cognitive radio node in CRN, as shown in table 1, can be considered as an important factor that makes MAS very suitable to resolve cognitive radio issues. Before discussing the utilization of MAS in CRN (section IV), we will first introduce multi-agent concept in the following section.

Table 1. Comparison between an agent and a cognitive radio node

Agent	Cognitive radio node
- An agent is a virtual entity that can perceive its environment, act and communicate with other agents.	- A cognitive radio terminal interacts with its radio environment, detects the free frequencies and then exploits them.
- Agent is autonomous and has skills to achieve its goal.	- The cognitive radio node has enough capabilities allowing it to manage the radio resources.

3. MULTI-AGENT SYSTEMS REVIEW

Since its introduction in 1956, the artificial intelligence (AI) was a branch of computer science that focuses on machine intelligence. The objective is to produce a system simulating human reasoning, considering only one actor to solve problems.

The idea of Distributed Artificial Intelligence (DAI) is to move from individual to collective behavior in order to address the limitations of traditional AI when solving complex problems requiring the distribution of intelligence over several entities. The DAI includes three basic research areas which are: distributed problem solving, parallel artificial intelligence and multi-agent systems.

In this paper, we focus on MAS and their application in cognitive radio networks as a solution for a more efficient spectrum management and radio resource allocation. The next sub-section describes main issues of multi-agent systems.

3.1. Concept and Definitions

An MAS [24] consists in multiple interacting computing elements, known as agents. Agents are programs that can decide for themselves what they need to do in order to satisfy their design objectives. Besides, they are capable of interacting with other agents by engaging in analogous social activity types such *cooperation, coordination* and *negotiation* [24]. Ferber [25] gives a definition of an agent and an MAS. According to him, an agent can be defined as a physical or virtual entity that can act, perceive its environment and communicate with others; it is autonomous and has skills to achieve its goals. Besides, an MAS is a set of agents interacting in a common environment [25].

To be intelligent, an agent must have some properties such as reactivity, proactivity and social ability [26]. We detail these capabilities in the following sub-section.

3.2. Agent's properties

Three main features are required to make an agent intelligent [24, 26], which are:
1) Reactivity: Intelligent agents are able to perceive their environment and respond in a timely fashion to changes that occur in it in order to satisfy their objectives.
2) Proactivity: Intelligent agents are able to exhibit goal-directed behaviour by taking the initiative to reach their goal.
3) Social ability: Intelligent agents are capable of interacting with other agents (and possibly humans) in order to reach their aims.

Each agent may face difficulties to solve complex problems alone. Consequently, the intelligence is distributed among various components, which can communicate and cooperate with each other to realize their goals. This is the origin of MAS idea.

As we have mentioned, one of the main characteristics of an MAS is its social ability, i.e., its capacity of interaction and communication between agents existing in the same environment. Therefore, we discuss, next, different techniques of agent's interaction.

3.3. Agent's interaction mechanisms

Interaction and communication are often confused in the literature. In fact, communication is the transmission of information between agents while interaction contains two elements: communication and the actions resulting from the information exchange. Two types of communication can be distinguished: direct communication that consists in message exchange between agents and indirect communication that consists in signals or indicators transmitted through the environment as pheromones, for instance. We propose to classify agent's interaction into three categories: *coordination*, *cooperation* and *negotiation*. This classification is inspired from [24] and [27]. In Figure 2, we give different mechanisms and communication types related to each interaction class. In the following paragraph, we describe briefly these three types of agent's interaction.

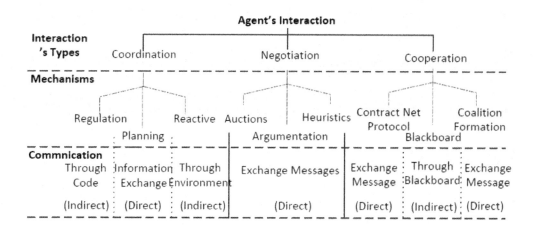

Figure. 2 Agents' interaction types

1) *Coordination:* It is a mechanism for ensuring that agent's activities preserve some desired relationships such as sequence and complementarity. The most popular coordination mechanisms are regulation, planning and reactive cooperation such as the one providing by the subsumption architecture [28]. The subsumption architecture is mainly used for ensuring reactive robot behavior. It is a way of decomposing complicated intelligent behavior into

many simple behavior modules, which are in turn organized into layers. Each layer implements a particular goal of the agent. The communication between agents in such architecture is indirect and can be made through the environment. The coordination by regulation is done through a communication based on a set of defined rules, for example by the use of code, law or social conventions. On the other hand, planning coordination defines a set of goals and plans and ensures the synchronization of agents' actions. This type of coordination can be centralized or distributed. In centralized coordination, a central agent prepares plans and distributes them to agents, solving synchronization and resource allocation problems. In distributed coordination, agents that can be planners and executors communicate their partial plans until goals are satisfied.

2) *Cooperation:* Cooperative agents work together to maximize theirs utilities and reach a common goal. In general, an agent cooperates with the others when it is not able to accomplish alone its task or if others are more efficient than it. As examples of cooperative approach, we can cite Blackboard [29], Contract Net Protocol (CNP) [30] and Coalition Formation [31]. The communication between agents in Blackboard technique is done via a shared blackboard where agents post their information. However, the Coalition Formation and the CNP are based on exchanging messages between agents.

3) *Negotiation:* It is a process that enables agents to reach agreement about the provision of a service by an agent for another one. Agents exchange messages in order to discuss their respective points of view. Negotiation can be used to resolve conflicts between agents. Different negotiation methods are proposed in the literature such as auctions [32], argumentation [33] and heuristics [34].

Many researchers mix coordination and cooperation mechanisms in a same category. That is why we can find CNP, blackboard and Coalition Formation defined as coordination mechanisms. Generally, reactive coordination is not highly efficient and planning and regulation mechanisms are very complex to implement. Therefore, planning, regulation and reactive coordination are rarely applied in cognitive radio networks. In the rest of the paper, we will focus on the most popular works using cooperation and negotiation mechanisms. Therefore, in the following two subsections, we will describe in detail multi agents' cooperation and negotiation protocols.

3.4. Cooperation Based Protocols

We will detail in this subsection Blackboard system, Contract Net protocol and Coalition Formation as cooperative MAS protocols.

3.4.1 Blackboard

The problem to solve is drawn up on a centralized blackboard and each participating agent contributes with its own knowledge until a sufficient solution is reached. Basically, a blackboard system [29] consists of three parts: (1) Knowledge Sources (KS) that provide the specific expertise needed by the application; (2) a blackboard, which contains the information of partial solutions; (3) and a control shell to maintain the coherence between various knowledge sources. All these parts work together to solve the assigned problems. This separation between knowledge (agents), space solution (blackboard) and knowledge management (control) gives the system a very high modularity and an easy access to knowledge. Besides, Blackboard technique provides a control mechanism that try at every resolution step to choose the best knowledge source. However, even if the control is providing an efficiency gain, it burdens the processing which results in a lack of flexibility. In addition, the blackboard is a centralised approach that depends highly on the central blackboard.

3.4.2 Contract Net Protocol

The contract net protocol (CNP) [30] is a MAS task-sharing protocol, consisting in a collection of software agents that forms the *'contract network'*. Each node of the network can, at different times or for different tasks, be a manager or a contractor based on its requirements. The manager is responsible for initiating the task and then monitoring its execution by exchanging a series of messages with the contractors while the contractor executes the assigned tasks. The CNP provides the advantage of real-time information and messages exchange making it suitable for situations where control and resources are distributed. CNP is fast and flexible. It is also used for task decomposition. However, when task is very complex and cannot be further decomposed into subtasks, CNP cannot be used.

3.4.3 Coalition Formation

Many tasks must be completed by more than one agent as its realization needs resources and capabilities that are beyond those of an agent alone. Very often, even if a task may be completed by a single agent, its performance can be too low to be acceptable. In such a situation agents may form groups to solve the problem by cooperation. Cooperative agents work together on a given task. Initially agents are independent and do not cooperate. When they cannot complete their tasks individually, agents may exchange information and try to form coalitions, which give them best efficiency [31]. Generally, in MAS Coalition Formation, the agents work in order to maximize the utility of the whole system, and after a successful completion of the assigned task the gained profit will be distributed equally or according to each agent contribution. Moreover, Coalition Formation makes possible the resolution of complex problem without task decomposition. However, this protocol is relatively slow due to the overhead resulting from the construction of coalitions.

3.5 Negotiation Based Protocols

In this subsection, we will detail the most popular negotiation based MAS mechanisms: auctions, argumentation and heuristics.

3.5.1 Auctions

An auction [32] takes place between an agent known as the auctioneer and a collection of agents known as bidders. The goal of auction is for the auctioneer to allocate the good to one of the bidders. Traditionally, four types of auctions are used: First-price sealed-bid auction (FPSB), Second-price sealed-bid auctions (Vickrey), Open Ascending-bid auctions (English auctions) and Open Descending-bid auctions (Dutch auctions).

In FPSB auction, bidders place their bid in a sealed envelope and simultaneously submit to the auctioneer. The envelopes are opened and the individual with the highest bid wins, paying a price equal to the exact amount bided. Vickrey auction is like FPSB auction but the winner pays a price equal to the second highest bid.

In English auctions, the price is steadily raised by the auctioneer with bidders dropping out once the price becomes too high. This continues until there remains only one bidder who wins the auction at the current price.

In Dutch auctions, the price starts at a level sufficiently high and is progressively lowered until a bidder indicates that he is prepared to buy at the current price. The winner pays the price at which he proposes.

We can find an hybrid form of auction called double auctions where participants are buyers and sellers in the same time and trade on the same product.

In almost all traditional settings except Vickrey, the auctioneer desires to maximize the price at which the good is allocated, while bidders desire to minimize the price. Auctions can be an effective solution to resolve conflicts between agents. Nevertheless, the major problems of auctions are the fraud and when the correct winner cannot be determined.

3.5.2 Argumentation

Argumentation [33] is the process of attempting to agree about what to believe by supporting arguments. It provides principles and techniques for resolving inconsistency or at least sensible rules for deciding what to believe in the face of inconsistency. Argumentation takes into account the bounded resources nature of real agents, by providing the possibility of acting before the completion of the reasoning process on the basis of provisional conclusions. However, this mechanism is very complex and need knowledge about past actions and agents' arguments.

3.5.3 Heuristics

Heuristics [34, 35] are mathematical and learning based solutions. For example, in a competitive market, the buyer agents can exchange false bids to increase their payoffs. Using heuristics, the seller agents can learn from the previous bids and information exchanges of malicious buyers and hence, they can decide to avoid any future trades with the malicious buyers. From simple negotiation, agents can accept or refuse proposals from other agents that can make the negotiation very long and inefficient especially when agents do not understand the cause of reject. Heuristics can also improve negotiation efficiency since agents can provide more useful feedbacks about the received proposals. These reactions may take the form of critique or cons/modified proposal. The difficulty in this technique is the choice of the more efficient heuristic.

In the next section, we will present the utilization of some of these different MAS protocols for the resolution of cognitive radio spectrum management issues.

4. MULTI-AGENT SYSTEMS APPLICATION IN COGNITIVE RADIO NETWORKS

MAS are relatively a new, yet, one of the most popular concepts in the research community. It is widely applied in the telecommunications and network domains. The involvement of MAS in the new research area of cognitive radio is notably present.

Agent's intelligence and cognition are important features in a cognitive radio network as it can help to perceive the environment and react properly. In addition, agent's interaction protocols such as negotiation and cooperation can provide more effective communication between network entities and can lead to a better exploitation of unused spectrum portions.

Recently, a large range of studies have used MAS to ensure efficient utilization of available spectrum resources in CRN. We propose to classify the existing works that apply MAS mechanisms for the resolution of cognitive radio problems into three main categories: negotiation, learning and cooperative based approaches, as shown in Figure 3.

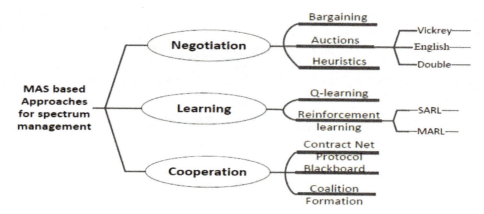

Figure. 3 Three main categories of MAS based approaches for spectrum management in CRN

4.1. Negotiation Based Approaches

Negotiation based approaches are largely used to address spectrum allocation issue in cognitive radio network especially auctions based ones. Agents' negotiation is a way to maximize user's utility since PUs and SUs negotiate in order to reach an agreement that best suits their needs. Cognitive radio solutions using multi-agent negotiation can be classified into three classes according to the adopted negotiation protocol: bargaining, heuristics and auctions.

Bargaining is a type of negotiation mechanism based on selling and buying concepts. Bargaining has been used in CRN to solve sensing issue [37] or to provide fair spectrum allocation [38, 39]. In [37], bargaining based pairwise cooperative spectrum sensing scheme is proposed. According to their locations regarding the PU, SUs are grouped into pairs. In the cooperative sensing pairs, the nearest SU to the PU senses the band and relays the detected information to the other CR users. In each pair, rational SUs can bargain with each other over the sensing time division, and thus save the sensing time for data transmission. Results show that this proposal improves the sensing accuracy. However, the distance between CR users and the PU is considered as the only factor affecting the bargaining results. In addition, users are assumed to be identical and their movement is simplified.

A distributed spectrum allocation solution via local bargaining is presented in [39] where SUs self-organize into small bargaining groups. The group formation process starts by sending a group formation request for a subset of spectrum portions from an initiator SU to the neighboring SUs. The interested SUs answer the request and a bargaining group is formed. Once the group is formed, users adapt their spectrum assignment to approximate a new optimal assignment. The group is later dismissed at the arrival of a PU or after completely utilizing the channels for the agreed time period. In such solution, network nodes are supposed to be collaborative but in the real system they can be selfish so that a pricing based bargaining would be more realistic.

Besides, ***heuristics*** based solutions in cognitive radio context are provided to solve complex problems in spectrum allocation issue. For example in [40], the problem of allocating the channels among cognitive radio nodes is handled. The proposal uses graph theoretic techniques to find an optimal and valid allocation when channels are limited. As the problem is an *NP* problem, authors have used novel distributed heuristic that relies on a local Common Control Channel. It is proved that this heuristic outperforms those existing in the literature and it can be easily implemented. Yet, heuristics techniques are rarely used in CRN because they are complex to define.

Moreover, *auctions* are popular negotiation processes used in CRN in order to solve and optimize spectrum utilization task. In such solutions [41-46], all users submit their bids to a centralized manager. Auctioneer allocates the resources in a way that maximizes its utility. The utility function changes according to auctions type (English, Vickrey, etc.).

In [41], a spectrum management policy based on the Vickrey auction (*multi- unit sealed bid*) is proposed. Cognitive radio mobile stations can compete for the utilization of the available PUs spectrum bands. SUs (bidders) submit bids without knowing bids of the other SUs in the auction, and the highest bidder wins, but the price paid is the second-highest one (Vickrey auction). A pricing and billing based solution is proposed in [42] presenting an auction sequence mechanism, which allow users to express their urgency, needs, purchase power and preferences. Vikrey auction [41] seems to be more suitable to execute in a given time when compared to English auction used in [42].

Double auction based spectrum trading scheme [44] has been proposed to resolve the spectrum access between PUs and SUs. Two different utility functions for PUs and SUs are designed based on a supply and demand relationship between them. Buyers (PUs) submit their bids and potential sellers (SUs) simultaneously submit their prices. Different from the traditional spectrum sharing approaches, in [43] proposal's, the SUs are allowed to make decision simultaneously and independently and make bid decisions based on self-interested considerations. Simulation results show the effectiveness of this spectrum trading methodology.

Another auction based approach is proposed in [47] by considering a cognitive radio network consisting of one PU and multiple SUs sharing the same spectrum. Each SU makes a bid for the amount of spectrum it requires and the PU may assign the spectrum band to the SU that do not degreed the PU's quality of service (QoS).

The spectrum trading model proposed in [48] uses agent as an additional part in the trading process that is different from secondary and primary users. The proposed trading model maximizes the payoff of the agent as well as it enhances the satisfaction of cognitive radio users. This model is flexible and easy to implement on the existing infrastructures. Authors give the example of an IEEE 802.22 network where agents are built on base stations that sense the VHF/UHF TV bands and serve all their associated clients without harmful interference to TV receivers.

4.2 Learning Based Approaches

Learning methods have been recently used to solve multiple issues in CRN. As examples, we can mention [49-51] for spectrum sensing, [52, 53] for spectrum sharing and [54] for spectrum access and control. We can distinguish two types of learning algorithms generally used in this context: reinforcement learning and Q-learning.

Reinforcement learning (RL) allows agents to learn from their past states (sharing pattern, neighborhood movement, etc.) in order to better perform their following actions and moves. In [49], authors propose a multi-agent learning approach, where each SU senses its nearby spectrum, perceives its current transmission parameters and takes necessary actions when PU appears. Several penalty values are enforced when the SU agents try to interfere with each other or with the PUs. This approach proves that allowing secondary users learning via interaction can improve the overall spectrum usage especially for the SUs' transmissions and channels switching capabilities. Two types of RL approaches are discussed in [56], namely Single Agent Reinforcement Learning (SARL) and Multi-Agent Reinforcement Learning (MARL). SARL has been applied in an operating environment with a single agent such as the base station in a centralized network, while MARL has been applied in an operating environment with multiple agents, such as all the SUs in a dynamic cognitive radio network. In MARL solution, agents learn and take their own respective actions, in a cooperative and distributed manner, as part of the joint action that maximizes the overall network performance. The difference between SARL and MARL is the Payoff Message Exchange (PME) that is an additional feature in MARL. The PME is the mean of communication for the learning engine embedded in each agent. SARL does not implement PME because it is a single-agent approach. In general, as proved in [56], RL provides high network-wide performance with respect to dynamic channels selection. However, they are mostly difficult to realize. For this reason, a recent form of RL that makes agent learning process easier has emerged, known as Q-learning.

Q-learning is an out part of reinforcement learning method that does not need a model of its environment. Q-learning algorithm is increasingly used for spectrum access because it allows agents to learn easily how to act optimally. For example, in [54], each SU senses channels and then chooses an unused frequency channel to transmit data unless other SUs exist. If two or more secondary users choose the same channel for data transmission, collision occurs. Such a scheme is similar to Aloha protocol where no explicit collision avoidance is applied. However, the SUs can try to learn collision avoidance, as well as channel qualities, according to their experience. Q-learning is exploited in [55] to ensure channels selection in CRNs. In [55], SUs periodically share the relative traffic information on the sensed channels with their neighbors. Based on this information exchange, the SUs dynamically tune their transmission powers on selected channels and avoid interferences. However, getting precise information in such learning schema can become more difficult if users have weak information about their neighbors' spectrum usages.

Learning methods show interesting efficiency in the resolution of cognitive radio tasks. They are increasingly used in the literature and still studied in order to provide more enhancements in the performance of cognitive radio systems. However, they are still faced to the high cost of learning and the uncertainty of its outcomes (i.e., the accuracy of what is learned is not guaranteed).

4.3 Cooperative Based Approaches

Some researchers have used specified MAS cooperation methods such as CNP, Coalition Formation and Blackboard to ensure efficient spectrum resource allocation.

A *Contract Net Protocol* based approach for spectrum sharing in CRN is proposed in [57]. This approach relies on Call for Proposal (CFP) concept. PU agent is considered as a "manager" and SUs are considered as "contractors". Each SU sends a CFP to PUs who are interested to share their spectrum bands. Based on a defined function, the SU characterizes each received proposal. Once the deadline to receive proposals expires, the SU sends an accept message to the PU that gives the best proposal and sends a reject message to all others. This cooperative framework enables cognitive radio devices to work cooperatively with their neighboring licensed devices in order to utilize efficiently the available spectrum dynamically. This technique enhances the overall spectrum allocation and ensures users satisfaction. However, this method is provided to solve the problem of allocation only between PUs and SUs and not consider spectrum sharing problem between SUs when no PU is occupying the band.

Coalition Formation is also used in [58] for sharing unlicensed spectrum bands. The band is assumed to be occupied by only secondary users. SUs are equipped with agents and interact with their neighbors to form several coalitions over the unlicensed bands. These types of coalitions can provide a less-conflicted spectrum access as the agents cooperatively agree for spectrum sharing. Another goal of using multi-agent coalition formation is to create cohesive groups that benefit from agents' participation in terms of their mutual successful access.

Recently, a new architecture for cognitive spectrum management based on cooperative agents has been proposed in [59]. In this approach, intelligent agents that are embedded in the radio devices coordinate their operations to exploit network's resources and avoid interference with PUs. Agents carry a set of modules to gather information about the terminal status and the radio environment and act according to the constraints of the user's application. Cognitive radio nodes that wish to transmit data will rely on their neighborhood agents' information to determine the status of the spectrum occupation. Coalition Formation is mainly used to reduce computational efforts on the cognitive radio terminal when collecting its environmental information.

Blackboard based system is adopted in [60] to present a cross-layer architecture for cognitive radio networks. The goal of this new architecture is to ensure efficient spectrum allocation and to improve overall quality of service. Each cognitive node generates parameters related to the current network state and deposes it on the blackboard in order to optimize bandwidth utilization. This blackboard approach helps to solve some related network problems such as intrusion detection and jamming.

As previously described, most of the studies based on multi-agent concept in cognitive radio networks focuses on spectrum sensing and allocation. However, no much MAS based research is made on the topic of spectrum mobility in CRN. Very few works have used cognitive radio and MAS concepts to improve mobility management in traditional cellular networks. For example, the proposed approach in [61] enables modification in base stations' parameters to meet new services requirements. These changes are performed using agents that manage cells via negotiation, learning and reasoning strategies. In [61], author's main goal was to reduce interference, handoff delay and blocking probability.

In [62], a proposal to solve the problems of spectrum mobility, sharing and handoff in CRN using MAS is presented. A mobile cognitive radio network is considered where each terminal is managed by an agent. Authors propose an algorithm that should be executed by a mobile cognitive radio terminal when moving from one geographical zone to another one. When the mobile cognitive radio user comes close to a new zone, it tries to collect information about its new environment in order to anticipate a possible handover. According to the recorded information, the CR user updates its knowledge base with, among others, spectrum conditions and PUs preferences such as price and use duration for the potentially allocated band. A possible negotiation process may be activated between the SU and the PU having the channel that best suits the user's requirements.

From the above discussion, we can confirm that multi-agent systems, which are variously and largely used for spectrum management in CR networks, can provide very efficient solutions to many cognitive radio issues. Using MAS in cognitive radio networks is promising and presents and open research area that can be more explored.

5. Conclusions

Cognitive radio is a promising technology that plays an important role in the exploitation of the existing spectrum resources. It seriously participates in enhancing the spectrum utilization by allowing opportunistic access to spectrum holes.

In this survey, we provided first a presentation of cognitive radio paradigm. Then, we gave an overview of multi-agent systems concept and their application in cognitive radio networks. We classify MAS based researches within CRN in three categories: negotiation mechanisms, learning methods and cooperative approaches. For each category, we explained the fundamental concepts and we provided examples from the literature.

In addition to existing game theory based approaches to solve CRN issues, most contributions using MAS in CRNs are based on auctions mechanisms. Learning methods have been recently used and they are evolving. However, cooperative approaches are not widely applied and still represent an open research topic in cognitive radio domain that may require more studies and investigation of novel proposals.

ACKNOWLEDGEMENTS

This work is partly supported by the Ministry of Higher Education and Research of France.

REFERENCES

[1] P. Ren, Y. Wang, Q. Du and J. Xu (2012), "A survey on dynamic spectrum access protocols for distributed cognitive wireless networks", *EURASIP Journal on Wireless Communications and Networking*.

[2] J. Mitola (2006), "*Cognitive radio architecture : The Engineering Foundations of Radio XML Link*", John Wiley and Sons.

[3] I.F. Akyildiz, W. Y. Lee, and K. R. Chowdhury (2009), "CRAHNs: Cognitive Radio Ad Hoc Networks," *Ad Hoc Networks*, vol.7, pp.810-836.

[4] D.C Oh, Y-H. Lee (2009), "Energy Detection Based Spectrum Sensing for Sensing Error Minimization in Cognitive Radio Networks", International Journal of Communication Networks and Information Security, vol. 1, no. 1, pp. 1-5.

[5] Z. Yaqin, L. Shuying, Z. Nan and W. Zhilu (2010), "A Novel Energy Detection Algorithm for Spectrum Sensing in Cognitive Radio", *Information Technology Journal*, vol 9*, pp.* 1659-1664.

[6] A. Ghasemi, and E.S. Sousa (2005), "Collaborative spectrum sensing for opportunistic access in fading environment", *IEEE Symposia on New Frontiers in Dynamic Spectrum Access Network*, pp.131- 136.

[7] M. E. Sahin, I. Guvenc and, H. Arslan (2010), "Uplink user signal separation for OFDMA-based cognitive radios" *EURASIP, Journal on Wireless Communications and Networking*.

[8] A. Bouzegzi, P. Ciblat and P. Jallon (2008), "Matched Filter Based Algorithm for Blind Recognition of OFDM Systems", *IEEE VTC-Fall*, pp. 1-5.

[9] S. Kapoor, S.V.R.K. Rao, and, G.Singh (2011), "Opportunisitic Spectrum Sensing by Employing Matched Filter in Cognitive Radio Network", *CSNT*, pp. 580-583.

[10] B. Wang, Y. Wu and K. J. R. Liu (2010), "Game theory for cognitive radio networks: An overview", *Computer Networking*, vol. 54, n°14, pp. 2537-2561.

[11] J. Rajasekharan, J. Eriksson, V. Koivunen (2010), "Cooperative game-theoretic modeling for spectrum sensing in cognitive radios", *Conference Record of the Forty Fourth Asilomar Conference on Signals, Systems and Computers*, pp. 165-169.

[12] W. Saad, H. Zhu, T. Basar, M. Debbah, A. Hjorungnes (2011), "Coalition Formation Games for Collaborative Spectrum Sensing", *IEEE Transaction on Vehicular Technology*, pp.276-297.

[13] J. Rajasekharan, J. Eriksson, V. Koivunen (2011), "Cooperative Game-Theoretic Approach to Spectrum Sharing in Cognitive Radios", *Computer Science and Game Theory*.

[14] M. Liang, Z. Qi (2011), "An Improved Game-theoretic Spectrum Sharing in Cognitive Radio Systems",*3rd International Conference on Communications and Mobile Computing*, pp. 270-273.

[15] A. De Domenico, E. Calvanese Strinati, M.G Di Benedetto (*2012*), "A Survey on MAC Strategies for Cognitive Radio Networks", *IEEE Communications Surveys & Tutorials*, vol. 14, n°1, pp.21-44.

[16] X. Zhang, S. Hang (2011), "CREAM-MAC: Cognitive Radio-EnAbled Multi-Channel MAC Protocol Over Dynamic Spectrum Access Networks", *IEEE Journal of Selected Topics in Signal Processing*, vol. 5, n°1, pp.110-123.

[17] C.-W. Wang and L.-C. Wang (2012), "Analysis of Reactive Spectrum Handoff in cognitive radio networks", *IEEE Journal on Selected Areas in Communication,* vol.30, n°10, pp. 2016-2028.

[18] C. W. Wang, L-C. Wang, F. Adachi (2010), "Modeling and Analysis for Reactive-decision Spectrum Handoff in Cognitive Radio Networks", *IEEE Globecom*, pp.1-6.

[19] C. W. Wang, L-C. Wang (2009) , "Modeling and Analysis for Proactive-decision Spectrum Handoff in Cognitive Radio Networks", *IEEE International Conference on Communication (ICC), June*, pp. 1-6.

[20] Y. Song and J. Xie (2010),"Common Hopping based Proactive Spectrum Handoff in Cognitive radio Ad Hoc Networks, *IEEE Globcom*, pp. 1-5.

[21] J. Duan, Y. Li (2011), "An optimal Spectrum Handoff scheme for Cognitive radio mobile Adhoc Networks", *Advances in Electrical and Computer Engineering Journal*, vol. 11, n°3, pp. 11-16.

[22] L. Wang, L. Cao and H. Zheng (2008), "Proactive channel access in dynamic spectrum networks", *Physical Communication*, vol. 1, n°2, pp. 103-111.

[23] Y. Song, J. Xie (2012), "ProSpect: A Proactive Spectrum Handoff Framework for Cognitive Radio Ad Hoc Networks without Common Control Channel", *IEEE Transactions On Mobile Computing,* vol. 11, n°7, pp. 1127-1139.

[24] Michael Wooldridge (2002),"*An Introduction to Multi-Agent Systems*", Wiley and Sons Editor, West Sussex, England.

[25] J. Ferber (1999), "*Multi-Agent System: An Introduction to Distributed Artificial Intelligence*", Addison Wesley Longman.

[26] G. Weiss (2000), "*A modern approach to distributed artificial intelligence*", MIT press, USA.

[27] B. Espinasse, S. Fournier, F. L. Gonçalves de Freitas (2008), "Agent and ontology based information gathering on restricted web domains with AGATHE", *ACM symposium on Applied Computing*.

[28] M. Ahmed, V. Kolar, M. Petrova, P. Mahonen, S. Hailes (2009), "A component-based architecture for cognitive radio ressource management", *International Conference on Cognitive Radio Oriented Wireless Networks and Communications (CRWNCOM'09),* pp. 1-6.

[29] F. Jurado, M-A. Redondo, M. Ortega (2012), "Blackboard architecture to integrate components and agents in heterogeneous distributed eLearning systems: An application for learning to program", *Journal of Systels and Software*, vol. 85, n°7, pp. 1621-1636.

[30] U. Mir, S. Aknine, L. Merghem-Boulahia, and D. Gaïti (2010), "Agents' Coordination in Ad-hoc Networks," *8th ACS/IEEE International Conference on Computer Systems and Applications (AICCSA'10)*, Tunisia pp. 1-8.

[31] W. Gruszczyk, H. Kwasnicka (2008), "Coalition Formation in multi-agent systems- an evolutionary aproach", *International Multiconference on Computer Science and Information Technology (IMCSIT)*, pp. 125-130.

[32] G. Adomavicius and A. Gupta (2005), "Toward comprehensive real-time bidder support in iterative combinational auctions", *Information Systems Research*, vol. 16, n°2 .

[33] I. Rahwan, S.D Ramchurn and N.R Jennings (2003), "Argumentation-based negotiation", *The Knowledge Engineering Review*, vol. 18, pp.343-375.

[34] A. Amanna, D. Ali, D. Gonzalez Fitch, J. H. Reed (2012), "Hybrid Experiential-Heuristic Cognitive Radio Engine Architecture and Implementation". *Journal of Computer Networks and Communications,* 15 pages.

[35] D. Hu, S. Mao (2012), "Cooperative relay interference alignment for video over cognitive radio networks", *IEEE INFOCOM*, pp. 2014-2022.

[36] J. Xie, I. Howitt and A. Raja (2007), "Cognitive radio resource management using multi-agent systems", *Consumer Communications and Networking Conference (CCNC)*, pp.1123-1127.

[37] M. Pan, Y. Fang (2008), "Bargaining based pairwise cooperative spectrum sensing for Cognitive Radio networks", *IEEE Military Communications Conference*, pp. 1-7.

[38] H. Liu, A.B MacKenzie, B. Krishnamachari (2009), "Bargaining to Improve Channel Sharing between Selfish Cognitive Radios", *IEEE Globecom*, pp. 1-7.

[39] L. Cao and H. Zheng (2005), "Distributed spectrum allocation via local bargaining", *IEEE Communications Society Conference on Sensor and Ad Hoc Communications and Networks*, pp. 475-486.

[40] S. Vijay, R. Rao, V. Prasad, C. Yadati, and I.G Niemegeers (2010), "Distributed heuristics for allocating spectrum in CR ad hoc networks", *IEEE Globecom*, pp. 1-6.

[41] H.-B. Chang and K.-C. Chen (2010), "Auction-based spectrum management of cognitive radio networks," *IEEE Transactions on Vehicular Technology*, vol. 59, n°4, pp. 1923-1935.

[42] C. Kloeck, H. Jaekel, F. K. Jondral (2006), "Multi agent wireless System for Dynamic and Local Combined Pricing, allocation and Billing", *Journal of communication*, vol. 1, n°1, pp. 48-59.

[43] D. Niyato and E. Hossain (2008), "Competitive pricing for spectrum sharing in cognitive radio networks: Dynamic game, inefficiency of Nash equilibrium and collusion" *IEEE Journal on Selected Areas in Communication*, vol. 26, n°1, pp. 192-202.

[44] Y. Teng, Y. Zhang, C. Dai, F. Yang and M. Song (2011), " Dynamic spectrum sharing through double auction mechanism in cognitive radio network", *IEEE Wireless Communication and Networking Conference (WCNC)*, pp. 90-95.

[45] H. S Mohammadian, B. Abolhassan (2010), "Auction-based Spectrum Sharing for Multiple Primary and Secondary Users in Cognitive Radio Networks", *IEEE Sarnoff Symposium*, pp. 1-6.

[46] G.S Kasbekar, S. Sarkar (2010), "Spectrum auction framework for access allocation in cognitive radio networks", *IEEE/ACM Transactions on Networking*,vol. 18, n°6, pp. 1841-1854.

[47] X. Wang, Z. Li, P. Xu, Y.Xu, X. Gao, and H. Chen (2010), "Spectrum sharing in cognitive Radio Networks- An Auction based Approach", *IEEE Transactions on Systems, Man, and Cybernetics, Part B: Cybernetics - Special issue on game theory*, vol. 40, n°3, pp. 587-596.

[48] L. Qian, F. Ye, L. Gao, X. Gan, T. Chu, X. Tian, X. Wang, M. Guizani (2011), "Spectrum Trading in Cognitive Radio Networks: An Agent-Based Model under Demand Uncertainty", *IEEE Transactions on Communications*, vol. 59, n°11, pp. 3192-3203.

[49] C. Wu, K. Chowdhury, and M. D. Felice (2010), "Spectrum management of cognitive radio using multi-agent reinforcement learning," *International Conference on Autonomous Agents and Multiagent Systems*.

[50] J. Lunden, V. Koivunen, S.R Kulkarni, H.V Poor (2011), "Reinforcement learning based distributed multiagent sensing policy for cognitive radio networks", *IEEE DySPAN*, pp. 642-646.

[51] Ch. Zhe, R.C. Qiu (2011), "Cooperative spectrum sensing using Q-learning with experimental validation", *IEEE Southeastcon*, pp. 405-408.

[52] F. Fu and M. van der Schaar (2008), "Dynamic Spectrum Sharing Using Learning for Delay-Sensitive Applications", *International Conference on Communication (ICC)*.

[53] P. Venkatraman, B. Hamdaoui, B., and M. Guizani (2010), "Opportunistic bandwidth sharing through reinforcement Learning", *IEEE Transactions on Vehicular Technology*, vol. 59, n°6, pp. 3148- 3153.

[54] H. Li (2010), *"Multiagent Q -Learning for Aloha-Like Spectrum Access in Cognitive Radio Systems"*, *EURASIP, Journal on Wireless Communications and Networking*.

[55] H. Li (2009), "Multi-agent Q-Learning of Channel Selection in Multi-user Cognitive Radio Systems: A Two by Two Case", *IEEE Conference on System, Management and Cybernetics*, pp. 1893-1898.

[56] K-L-A Yau, P. Komisarczuk, and P.D. Teal (2010), "Enhancing Network Performance in Distributed Cognitive Radio Networks using Single-agent and Multiagent Reinforcement Learning", *IEEE Local Computer Networks (LCN)*, pp. 152-159.

[57] U. Mir, L. Merghem-Boulahia, and D. Gaïti (2010), "COMAS: A Cooperative Multiagent Architecture for Spectrum Sharing," *EURASIP, Journal on Wireless Communications and Networking*, 15 pages.

[58] U. Mir (2011), "Utilization of Cooperative Multi-agent Systems for Spectrum Sharing in Cognitive Radio Networks", Phd dissertation, University of technology of troyes.

[59] A. Ahmed, O. Sohaib, W. Hussain (2011), "An Agent Based Architecture for Cognitive Spectrum Management", *Australian Journal of Basic and Applied Sciences*.

[60] Y.B Reddy, C. Bullmaster (2008),"Cross-Layer Design in Wireless Cognitive Networks", *International Conference on Parallel and Distributed Computing, Applications and Technologies (PDCAT)*, pp. 462-467.

[61] J. Raiyn (2008), "Toward cognitive radio handover management", *IEEE 19th PIMRC*, p. 1-5.

[62] E. Trigui, M. Esseghir L. Merghem boulahia (2011), "Spectrum Access during Cognitive Radio Mobiles' Handoff", *International Conference on Wireless and Mobile Communications*, pp 221-224.

An Intercell Interference Coordination Scheme in LTE Downlink Networks based on User Priority and Fuzzy Logic System

A. Daeinabi[1], K. Sandrasegaran[1], and X.Zhu[2]

Centre for Real-time Information Networks,
School of Computing and Communications, Faculty of Engineering and Information
Technology, University of Technology Sydney, Sydney, Australia
Ameneh.Daeinabi@student.uts.edu.au,
Kumbesan.Sandrasegaran@uts.edu.au
School of Information and Communications, Beijing University of Posts and
Telecommunications Beijing, China
zhuxn@bupt.edu.cn

ABSTRACT

The Intercell Interference (ICI) problem is one of the main challenges in Long Term Evolution (LTE) downlink system. In order to deal with the ICI problem, this paper proposes a joint resource block and transmit power allocation scheme in LTE downlink networks. The proposed scheme is implemented in three phases: (1) the priority of users is calculated based on interference level, Quality of Service (QoS) and Head of Line (HoL) delay;(2) users in each cell are scheduled on the specified subbands based on their priority; and (3) eNodeBs dynamically control the transmit power using a fuzzy logic system and exchanging messages to each other. Simulation results demonstrate that the proposed priority scheme outperforms the existing Reuse Factor one (RF1) and Soft Frequency Reuse (SFR) schemes in terms of cell throughput, cell edge user throughput, delay and interference level.

KEYWORDS

LTE, intercell interference coordination, resource block allocation, transmit power allocation, fuzzy logic.

1. INTRODUCTION

In recent years, the demand for mobile broadband services with higher data rates and better Quality of Service (QoS) is growing rapidly and this demand has motivated 3GPP to work on Long Term Evolution (LTE). One of their main goals was to define a simple protocol which involves both the radio access network (RAN) and the network core [1]. Moreover, it can obtain the peak rates of 100 Mb/s and a radio network delay of less than 5 ms, improve the spectrum efficiency and support the flexible bandwidth. In addition, the new flat network architecture can reduce the latency rather than 3G [2]. The multiple access technologies on the air interface are different in downlink and uplink of LTE systems; Orthogonal Frequency Division Multiple Access (OFDMA) is the downlink multiple access technology while for uplink, Single Carrier Frequency Division Multiple Access (SC-FDMA) is deployed [3]. Moreover, LTE supports frequency division duplex (FDD), time division duplex (TDD) as well as the wide range of system bandwidths which enables the system to work in a great number of different spectrum allocations [4].

Since the radio bandwidth is one of the scarce resources in wireless networks, new resource allocation algorithms need to be introduced to overcome radio resource limitation particularly when applications with high data rate are deployed. For this purpose, frequency reuse one has been used in cellular networks. However, the system performance is severely degraded due to increase of interference caused by neighbouring cells. There are two major categories of interference for cellular mobile communication system: intracell interference and intercell interference. Since the LTE downlink systems use OFDMA [5], the orthogonality among subcarriers is designed to mitigate the intracell interference. However, the intercell interference (ICI) caused by using the same frequency in neighbouring cells, can restrict the LTE performance in terms of throughput and spectral efficiency, particularly for cell edge users (CEU). Note that CEUs are user equipments (UEs) in a cell which are far from the serving eNodeB (eNB) while cell centre users (CCUs) are close to the serving eNB. Therefore, the ICI mitigation is a critical point to improve the performance of the system.

In this paper, we consider a dynamic priority based scheme in which the resource allocation for each subband is performed based on user priority. In addition, the transmit power of each Resource Block (RB) is dynamically determined through a fuzzy logic system. Note that a RB is defined as a smallest radio resource which is allocated to a UE. In addition, the system traffic loads as well as system changes are taken into account when the RBs and transmit powers are allocated to different UEs. The proposed scheme could jointly optimize RB and power allocation for each cell, while in the traditional methods, the spectrum allocation or the power allocation is fixed in each cell. In addition, the proposed scheme is a decentralized scheme in which each eNB specifies its own RB allocation and transmit power by exchanging messages with neighbouring eNBs over X2 interface. Consequently, it could improve the system performance in terms of cell throughput and cell edge throughput.

The rest of the paper is organized as follow. The ICI formulation is provided in Section 2. In Section 3, the related work is reviewed. In Section 4, the proposed ICIC scheme is described. Simulation results are presented in Section 5. The conclusion is given in the final Section.

2. INTERCELL INTERFERENCE FORMULATION

When a UE moves away from the serving eNB and becomes closer to the neighbouring eNB, the strength of desired received signal decreases and the ICI increases. The impact of ICI in LTE downlink can be analysed using the received Signal to Interference and Noise Ratio (SINR) of UEm on RBn as given in (1).

$$SINR_{m,n} = \frac{p_n^l H_{m,n}^l}{\sum_{k \neq l} p_n^k H_{m,n}^k \delta_n^k + P_N} \quad (1)$$

where P_n^l and P_n^k are the transmit power from the serving cell l and neighbouring cell k on RBn, respectively. Moreover, $H_{m,n}^l$ and $H_{m,n}^k$ denote the channel gain from the cell l and cell k to UEm on RBn, and P_N is the noise power. δ_n^k is set to 1 or 0 to indicate whether the neighbouring cell k allocates RBn to its UEs or not. It can be summarised that three important factors have a significant influence on the SINR of each UE namely the channel gain from eNB to UE, transmit power on each RB and RB allocation scheme.

From (1), it can be observed that the SINR of UEs closer to their serving eNB is higher compared to the UEs further away from their serving eNB. Furthermore, UEs at the cell boundary would experience higher interference while the desired signal is relatively low. This situation greatly affects the SINR value and the supportable services to users at the cell edge area. As a result, in many ICI mitigation schemes, UEs are classified into CEU and CCU. For different UE

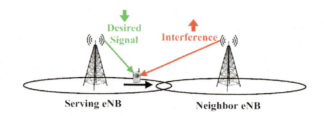

Figure 1. Illustration of UE moving away from its serving eNB

classifications (i.e. either CEU or CCU), alternative power and frequency allocation schemes can be adopted. This allows each UE, particularly CEU, increasing the transmit power which leads to a conflict over the system performance. More specifically, transmitting at higher power for CEU may marginally improve the SINR but results in a significant increase of overall interference in the system. Consequently, the higher transmit power impacts the actual overall system performance of all UEs. Therefore, in order to improve the system performance, a number of power allocation schemes are proposed to achieve an optimal compromise between the achieved SINR and the resulting interference. The third factor turns the ICI mitigation scheme into a variety of frequency or RB allocation algorithms in which the objective of optimization is to reduce the ICI and achieve higher spectral efficiency simultaneously, especially for the CEUs. From (1), it can be seen that the interference can significantly decrease when the serving cell transmits data on RBn if the neighbouring cells do not allocate the RBn to their UEs, (i.e.,) . The granularity of the allocation can be a RB or a portion of the available bandwidth. The concept to reserve parts of the bandwidth to avoid interference is classified as a frequency reuse technique.

3. RELATED WORK

Interference avoidance [6] is one of ICI mitigation strategies which deploys certain restrictions on the resources used in different cells such as restriction on frequency or power. In this section, we classify proposed avoidance algorithms into two classes: non-intercell coordination and Intercell coordination schemes and each of them can include different time scale schemes.

3.1. Non- Intercell Coordinated Schemes

Reuse factor one (RF1) is the simplest frequency reuse scheme in which the total available bandwidth is reused in each cell. In reuse factor three (RF3), the available bandwidth is divided into three equal orthogonal subbands and they are assigned to three neighbouring cells so that cells deploy different frequencies. In Softer Frequency Reuse (SerFR) scheme, the RF1 is used in cell edge and cell centre [7]. The main idea is to schedule RBs with more flexibility so that CEUs are allocated with a higher probability to frequency band with a higher transmit power and CCUs are assigned with a higher probability to frequency band with the lower transmit power. Some threshold values can be considered to separate CEU from CCU such as SINR, path loss, distance and so on.

In order to enhance the system performance of RF1 and RF3, the Fractional Frequency Reuse (FFR) techniques have been proposed. In FFR based schemes, the UEs who have higher signal quality use a lower reuse factor such as RF1 and UEs with the lower SINR deploy higher RF such as RF3 [8]. FFR techniques combine frequency and power allocation in different ways. Two well-known FFR schemes are Partial Frequency Reuse and Soft Frequency Reuse. In Partial Frequency Reuse (PFR) scheme [9], the total available bandwidth is divided into several subbands. For example, one subband is used for CCUs by using RF1 and three subbands are used for three neighbouring cells by using RF3 for CEUs. The power related to frequency of cell edge can be amplified while the total transmit power is fixed. For Soft Frequency Reuse (SFR) scheme ([10-12]), the one third of available bandwidth is allocated to CEU with amplified transmit power and the rest subbands are assigned to CCU with lower transmit power. The CEUs use only the cell edge subband while CCUs have access to whole available bandwidth with lower priority than CEUs. Also different types of SFR schemes have been suggested to enhance the system performance in [13-15]. Soft Fractional Frequency Reuse (SFFR) has been introduced in [16] to enhance the total cell throughput of FFR. SFFR scheme utilizes subbands allocated to the cell edge part in the neighbouring cells with low power levels for the CCU. The difference between SFFR and SFR is that the SFFR deploys the common subband.

Although the RF1 scheme has high peak data rate, the worst ICI will be observed in this scenario, particularly at cell edge parts. RF3 leads to low ICI improvement but the large capacity is lost because only one third of the resources are used in each cell. Therefore, the worst cell throughput is obtained by RF3. In PFR, The ICI is completely eliminated for the CEUs at the cost of the spectral efficiency losses. Although in SFR, UEs have higher spectral efficiency, they suffer higher ICI because the orthogonality between cell edge and cell centre subbands of neighbouring cells is not guaranteed. There are some analyses on FFR [17-19] and but [20] has provided an analytical comparison among RF1, RF3, PFR and SFR. It has been concluded that although static schemes enhance the cell edge throughput, they might decrease the cell throughput. Because of low throughput of RF1, RF3, PFR and SFR, other algorithms have been proposed to improve the performance of LTE downlink systems. Consequently, ICI mitigation cannot effectively improve the throughput only by the power control or resource limitation especially for those UEs who are close to each other in the system [21].Therefore, some algorithms have been developed to jointly perform on both resource and power allocations to maximize the throughput.

3.2. Intercell Coordinated Based Schemes

In coordination based schemes, eNBs should coordinate with each other to reduce ICI. Therefore, resource allocation is performed via information exchanged on X2 interface in the eNB level without using central entity. The coordination algorithms can be divided into two subclasses: fractional frequency reuse and dynamic frequency reuse schemes.

3.2.1. Fractional Frequency Reuse Schemes

In these schemes the dynamic thresholds are used for CCUs' RB allocation or dynamic power allocation. In Novel Enhanced Fractional Frequency Reuse (NEFFR) [22], the total available bandwidth is split into three parts where one part is allocated to cell edge with higher power and two parts are assigned to cell centre with lower power. An interference avoidance factor (IAF) based on fairness has been introduced to balance the number of UEs and RB allocations. Dynamic Fractional Frequency Reuse (DFFR) [23] divides entire available bandwidth into three subbands. The two first subbands have maximum power while the power of third subband is determined dynamically via an interference avoidance request (IAR) message. A semi dynamic ICI avoidance scheme has been proposed in [24] which combined the advantages of static coordination and dynamic coordination. According to non-overlapping handover regions, the CEUs in each cell will be divided into several cell edge groups. A dedicated RB list is assigned to each group. For cell edge group two steps are performed to achieve the list of RBs including the static resource pre-allocation and dynamic borrowing or leasing. Reference [25] introduced a region based ICI avoidance technique in which each cell is split into some areas according to the received interferences. To split the region, some UEs thresholds and UE measurements are used. Moreover, each region has its own specific power.

3.2.2. Dynamic Frequency Reuse Schemes

In these schemes, the resource allocation is performed based on optimizing a function or finding the priority of UEs.

One of the powerful methods for ICI avoidance is to optimize a function. Non-cooperation game theory is a method which has been used in [26] for optimizing. The utility function selects resources which receive the least interferences from neighbouring cells. A subsequent game theory approach has been suggested in [27] for real time systems in which the virtual token approach has been added into exponential rule. In [28, 29], a utility matrix is defined to cover all three sectors of an eNB including utility matrices for each sector and utility matrices for remaining sectors which use RBs simultaneously. These utility matrices are obtained based on the possible simultaneous ICI. A Hungarian algorithm [30] is executed on utility matrix which covers all three sectors and then, a threshold based approach is applied to determine which interferers should be restricted to maximize utility matrix. A subsequent autonomic ICI algorithm has been introduced in [31]. The proposed method considered the fuzzy reinforcement learning for learning and genetic algorithm (GA) for the solution of the ICI minimization problem using Relative Narrowband TX Power (RNTP) indicator. Graph based method is another methodology to optimize the desired functions. In [32], the problem of ICI has been mapped to the graph colouring problem.

Finding the priority of UEs based on their location or traffic models is a common approach to mitigate ICI particularly for CEUs. Using gain selective way is a different approach to determine UE which has the highest chance to receive a RB. This method has been deployed in [33] and can be performed by local search or total search. In [34], a QoS approach has been suggested for multiclass services. Since different traffics have different quality of service (QoS), resources should be assigned based on different QoS rather than using the same method. For latency-sensitive service, the [35] converts the QoS of a UE into the required number of RBs using size (bit) of packet, the remained scheduling time and channel capacity. Moreover, references [36-40] have reviewed some ICI mitigation algorithms proposed for LTE with more details.

Based on simulation results obtained from references, the schemes with intercell coordination have higher throughput than schemes without intercell coordination because in intercell coordinated based schemes, resource allocation is performed via information exchanged among eNBs. Moreover, among intercell coordination methods, dynamic frequency reuse schemes show better performance in terms of cell throughput and cell edge throughput because dynamic scheme can adapt instantaneously to network changes such as traffic variation or load distributions without using a priori resource partitioning. However, as a practical aspect, several factors should be considered when an algorithm would be selected as the best technique for a particular goal in real environment. There is usually a trade-off between optimal solutions on the one side and computational cost, complexity, time and overhead on the other side. When an ICI mitigation technique uses a complicated optimization method to find the best RB and power, although the system performance may increase significantly in theory, some new challenges will come up which impact on capability for real systems. For instance, high computational cost as well as high required overhead and time cause the system cannot trace the network changes and therefore cannot adapt rapidly. Consequently, the good theoretical algorithm may not be able to implement in the real world. In this paper we proposed a dynamic ICI mitigation scheme with simple fuzzy logic system to reduce the complexity.

4. THE PROPOSED ICIC SCHEME

In this scheme we focus on how to assign RB to different UEs and determine the transmit power of RBs in each cell. The main objective of this scheme is to mitigate the ICI in the LTE downlink system. Note that the downlink transmission is considered because the related broadband services pose higher rate requirements than those in uplink. The proposed scheme is executed in three phases as shown in Figure. 2. In phase A, we focus on how to find priorities of UEs and then sort the UEs based on their priorities. The subband related to each priority is determined in phase B. In phase C, the required transmit power for each RB is obtained through a fuzzy logic system.

Figure 2. Overview of the proposed algorithm

4.1. Phase A: Priority of UEs

In phase A, we find the priority of UEs both in cell l and its neighbouring cells which are going to use the same RB at the next Transmission Time Interval (TTI). When a UE comes away the serving eNB and becomes closer to a neighbouring eNB, the interference level receiving from the neighbouring eNB increases. Therefore, the UE should be scheduled on a RB which has higher transmit power in the serving cell and lower transmit power in neighbouring cells to reduce the ICI. Moreover, if different types of services are supported by the LTE system (e.g., real time and non-real time services), RBs should be allocated to different UEs based on their Quality of Service (QoS) requirements. Therefore, it is important that the QoS and Head of Line (HoL) are taken into account when the UE is selected to occupy one or more RBs. Consequently, we consider three parameters to determine the priority of UEs including interference level, QoS and HoL. Then, all parameters are normalized based on their maximum values and finally a weighting algorithm is deployed to find the final priority value.

4.1.1. Interference Level

In the first step, we calculate the impact of interference level, I_m, of UEm from neighbouring eNBs. Note that to calculate the interference level, an equal transmit power is considered for all eNBs on all RBs.

$$I_m = \sum_{k=1}^{K} p_m^n H_{m,n}^k \qquad (2)$$

In order to calculate the maximum interference, I_{max}, the following conditions is taken into account: 1) Minimum shadowing, 2) Minimum fading, 3) Maximum transmit power on the used RB, and 4) Minimum pathloss, that is when a UE is located on the boundary of two neighbouring cells, the maximum distance from the serving eNB and minimum distance from the neighbouring cell is obtained:

$$\text{Distance} = \frac{|Pos_l - Pos_k|}{2} \qquad (3)$$

where Pos_l and Pos_k are locations of serving eNB l and neighbouring eNB k. This assumption will lead to minimum path loss received from the neighbouring eNB. By replacing these values into (2), the I_{max} will be obtained.

4.1.2. Head of Line (HoL)

HoL represents the difference between the current time and the arrival time of a packet.

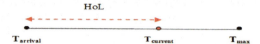

where the $T_{arrival}$ shows the time that the packet takes to arrive to the queue and T_{max} represents the maximum allowable packet delay for UEm.

4.1.3. Quality of Service (QoS)

QoS Class Identifier (QCI) is a scalar value considered as a reference to particular parameters that controls packet forwarding behaviours such as scheduling weights, admission thresholds, and queue management thresholds [41]. QCI is preconfigured by the operator owning the node (e.g. eNB). Note that the QCI is unique for each Service Data Flow (SDF). In addition, Guaranteed Bit Rate (GBR) is the minimum bit rate which applications request and usually deployed for applications such as Voice over Internet Protocol (VoIP). Non-GBR bearers cannot guarantee any specific bit rate, and are usually deployed for applications as web browsing.

Every QCI (GBR and Non-GBR) has a priority level in which the priority level 1 is the highest priority level. The priority levels would be deployed to differentiate between SDF aggregates of the same UE as well as SDF aggregates from different UEs [41]. We use the priority defined in Table 1 [41] as our third parameter and is represented by Q_m.

In order to combine the obtained parameters, we use a simple weighting algorithm shown in (4):

$$W = w_1 \times I'_m + w_2 \times HoL'_m + w_3 \times Q'_m \qquad (4)$$

$$\sum_{i=1}^{3} w_i = 1$$

Where w_i is a coefficient corresponding to each parameter. Parameters , , and represent the normalized values of the interference level, HoL and QCI priority of UEm, respectively. These normalized values are calculated as follows:

$$I'_m = \frac{I_m}{I_{max}} \qquad (5)$$

$$HoL'_m = \frac{HoL_m}{T_{max}} \qquad (6)$$

$$Q'_m = \frac{1}{Q_m} \qquad (7)$$

4.2. Phase B: Resource Block Allocation

The bandwidth division is based on UEs priority. The total available bandwidth BW is divided into three non-overlapping subbands shown in Figure 3 (a): 1) High Subband (*HS*): the subband which is allocated to UEs with higher priority; 2) low Subband (*LS*): the subband assigned to

Table 1. Standardized QCI characteristics [41]

QCI	Resource Type	Priority	Packet Delay Budget	Packet Error Loss Rate	Example Services
1	GBR	2	100 ms	10^{-2}	Conversational Voice
2		4	150 ms	10^{-3}	Conversational Video (Live Streaming)
3		3	50 ms	10^{-3}	Real Time Gaming
4		5	300 ms	10^{-6}	Non-Conversational Video (Buffered Streaming)
5	Non-GBR	1	100 ms	10^{-6}	IMS Signalling
6		6	300 ms	10^{-6}	Video (Buffered Streaming) TCP-based
7		7	100 ms	10^{-3}	Voice, Video (Live Streaming), Interactive Gaming
8		8	300 ms	10^{-6}	Video (Buffered Streaming) TCP-based
9		9			

UEs with lower priority, and 3) Common Subband (*CS*): the subband which can be allocated to all UEs with different types of priority. Therefore, RB allocation for UE with lower priority in *LS* does not affect the number of RBs that can be allocated to UEs with highest priority in *HS*.

After priority calculation, UEs in each cell as well as UEs in neighbouring cells which use the same RB should be sorted based on their priority. For this purpose, we first sort the UEs which are located in one cell. The UE with the highest priority is put in the first location of the queue. Then, the suitable RB from the relevant subband is allocated to each UE. In *HS*, the RB allocation is started by the UE with the highest priority while in *LS* the UE with the lowest priority is selected at first. Since the *LS* of one cell is selected as *HS* for another cell, it is possible that the UE in *HS* request the neighbouring eNB to reduce the transmit power on that RB in order to reduce the impacted interference. In addition, this approach allows us to use the remaining transmit power from *LS* for *HS* under the specified conditions to improve the throughput while the total transmit power is still smaller than or equal to the maximum transmit power of eNB. More details about transmit power will be described in power allocation.

4.3. Phase C: Power Allocation

Using a higher/lower transmit power by neighbouring eNBs on a same RB can directly lead to increase or decrease of UE's received signal levels. It will change the interference level impacted on UE by neighbouring eNB. The key idea of the proposed power allocation algorithm is that eNBs dynamically control the transmit power using fuzzy logic system (FLS) [42] and exchanging messages with each other. This algorithm can be executed for several TTIs until the power changes become small and after that it will be executed periodically or by a trigger.

After calculating the priority of UEs who are selected as a candidate to use a same RB (Figure.3 (b)), the transmit power of each eNB on that RB would be computed. For example, as we can see in Table 2, UE44 is located in *LS* of cell 1 while UE1 is in *HS* of cell 3. If the serving eNB of UE44 (eNB1) results highest interference on UE1 so that the SINR of UE1 becomes very low,

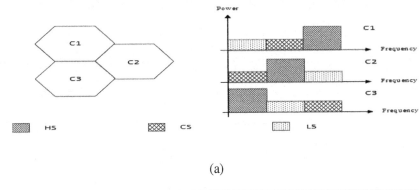

(a)

		Priority					
		Highest			----	-----	>Lowest
Queue	Cell 3	UE1	UE2	UE3	...	UE14	UE15
	Cell 2	UE16	UE17	UE18	...	UE29	UE30
	Cell1	UE31	UE32	UE33		UE43	UE44

(b)

Figure 3. (a) Bandwidth division, (b) locations of UEs in queues of different cells based on UE's priority

serving eNB of UE1 (eNB3) requests eNB1 to decrease its transmit power. In the other word, the strongest interferes have to decrease their transmit powers on a given RB if their UEs have lower priority on that RB while the minimum achievable data rate is satisfied. It is possible that one eNB has to decrease the transmit power of one RB several times, because its UEs has lower priority than UEs of neighbouring cells and it is detected as a strongest interferer for some neighbouring cells. In this case, the minimum obtained transmit power is selected as a transmit power for that RB. Moreover, if the total transmit power of eNB3 is smaller than the maximum transmit power, eNB3 can increase the transmit power on RB1. The amount of power decreasing or increasing could be obtained through FLS.

Table2. Allocating UEs to subbands for different cells based on UE's priority

Cell	RB1	RB2	...	$RB_{n/3}$	$RB_{(n/3)+1}$	$RB_{(n/3)+2}$	...	$RB_{(2n/3)}$	$RB_{(2n/3)+1}$	$RB_{(2n/3)+2}$	...	RB_{n-1}	RB_n
3	UE1	UE2	...	...	U15	U14	...	...	...	...	...	...	...
2	...	...	...	...	UE16	UE17	...	...	UE30	UE29	...	...	...
1	UE44	UE43	...	...	...	...	...	...	UE31	UE32	...	...	...
...	...	...	...	...	...	...	...	...	...	...	...	...	...

In order to determine the transmit power of each eNB on a particular RB, we use the FLS [42] because its design is simple and can work on time [43]. A FLS is an expert system using "IF... THEN" rules and includes very simple concept. It could simultaneously work with numerical data and linguistic information using a nonlinear mapping between input data and scalar output data. Since FLS deploys linguistic terms, the previous information can be gathered easier using the experience of an operator. The main difference between FLS and conventional rule base controller is that FLS can simultaneously trigger several rules which lead to smoother control. In general, FLS includes three steps shown in Figure 4:

1. *Input step*: it maps inputs to the appropriate membership functions and truth values.
2. *Processing step*: this step selects appropriate rules and finds their results. Then, it combines the results of the chosen rules.

3. *Output step*: in this step the obtained results are converted to a specific output value.

Through FLS, several states and actions could be defined. The SINR, historical throughput and achievable data rate are selected as the inputs of fuzzy interface because they are important metrics in LTE downlink system which can monitor the UE as well as system performances:

1) SINR of UE with higher priority.
2) Historical throughput of UE with lower priority.
3) The current achievable data rate that UE with lower priority could be achieved if it uses that RB. Each cell checks the obtained achievable data rate from its neighbouring cells through a data rate indicator. It is 2 if the achieved data rate is more than minimum achievable data rate, value 1 represents only the minimum achievable data rate is satisfied otherwise it is 0.

The crisp output of FLS is the transmit power of UEs which can change as follows:

1) Transmit power of UE with lower priority (i.e., *LS* or *CS*) could be fixed, avoided or decreased.
2) Transmit power of UE with higher priority (i.e., *HS*) might increase or be fixed.

In the inference stage, mapping of inputs to outputs are defined by a set of "IF-Then" rules. In the defuzzification stage, the output value is achieved using the aggregation of all rules and the centre

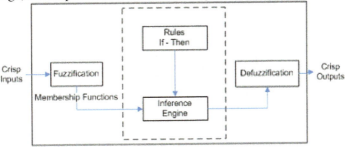

Figure 4. Overview of fuzzy logic system

of gravity approach. By implementation of FLS, the final transmit power for each RB in each cell would be obtained.

5. SIMULATION RESULTS

In order to evaluate the performance of the proposed priority scheme and compare with the well-known schemes, RF1 and SFR schemes, we describe the simulation results in this section.

5.1. Simulation Setup

Assessment is based on the system level simulation in a LTE downlink network with 19 cells. The network topology is composed of a set of cells and network nodes including macro eNBs and UEs which are distributed within cells. A certain area is defined for simulation where the eNBs and UEs are located and only in this region, UE movement and transmission are simulated. Each network node is identified by a unique ID and its position is defined using Cartesian system. A wraparound function should be used to make sure that the users do not exit from the simulation area. The simulation length is measured by TTIs which equals 1 ms. The simulation assumptions are outlined in Table 3. In addition, a streaming video is considered as the traffic model.

In order to consider power spectral density mask limitations, the power allocation is started based on SFR schemes. The initial transmit power of *HS* is presented by P_H and set to $\alpha \frac{P_T}{N}$ while the initial transmit power for both *LS*, P_L, and for *CS*, P_C, equal $\frac{(3-\alpha)}{2}\frac{P_T}{N}$ (α>1). Note that P_T is total transmit power of each cell and *N* is number of RBs. Moreover, a FLS has been implemented and a set of "IF-Then" rules defines the mapping of inputs to output including 27 rules for each output. For example "If SINR is low and the historical throughput is high and the achievable data rate is high Then the transmit power of *LS* reduces" or "If SINR is low and historical throughput is high and achievable data rate is high Then the transmit power of *HS* increases". The membership functions for inputs and outputs are specified as follows:

Table 3. Simulation parameters

Parameter	Assumption
Cellular layout	Hexagonal grid, 19 cells, wrap around
Inter-site distance	500 m
Carrier frequency	2 GHz
Carrier bandwidth	5 MHz
Channel model	Rayleigh
Shadow Fading	Log-normal
Path loss model	128.1+37.6 log(R), R in km
Total macro Tx power	43 dBm
UE speed	3km/h
Scheduler	Round Robin
α	2

1) Each input is fuzzified using three membership functions named "Low", "Medium" and "High" which are shown by "Z-shaped", "Triangular-shaped" and "S-shaped" respectively.
    - The range of membership functions is fixed for all SINR inputs according to SINR range obtained from SINR–Block Error Rate (BLER) curves [44].
    - The range of membership functions is fixed for all achievable data rate inputs according to Channel Quality Indicator (CQI) efficiency [44].
    - For throughput, the range of membership function could be changed based on the average number of RBs allocated to each UE in each cell.

2) To fuzzify the outputs, two different scenarios are assumed:
    - For transmit power of UE which belongs to *LS* and *CS*, the membership function of output is change from zero to maximum transmit power per RB. The membership functions are tagged with "Avoid", "Fix" and "Decrease".
    - For transmit power of UE in *HS*, membership functions are named by "Fix" and "Increase". The range of membership functions are changed from initial power to maximum transmit power plus to *ΔP*. Note that *ΔP* represents the additional transmit power which is obtained from reducing transmit power on other RBs for each cell.

Note that, in this paper the average cell throughput is defined as the amount of data sent successfully in a cell over a period of time (i.e., one TTI). Moreover, 5% edge throughput is considered as the 5th percentile point of the cumulative distribution function (CDF) of the user throughput which indicates the minimum throughput achieved by 95% of UEs [45].Cell edge user throughput means the average throughput of UEs located in cell edge regions. The results of the proposed priority scheme will be compared with RF1 and SFR schemes.

5.2. Performance Evaluation

As mentioned in Section 4, the RB allocation is based on a weighing algorithm with three coefficients including w_1, w_2 and w_3. Therefore, the difference in coefficient values can lead to different results as shown in Table 4. For example, when the coefficient value of interference level, w_1, is higher than w_2 and w_3, the interference level of UEs decreases which leads to improve the throughput. However, the delay increases because the UEs who have the higher interference level (i.e., CEUs) are selected as highest priority instead of UEs with larger delay. In this case, the interference level has higher priority than delay and type of service. On the other hands, when the highest coefficient value is allocated to delay, although the delay of system decreases, the throughput is reduced. Consequently, the coefficient values affect the system performance and they should precisely be specified.

In order to show the performance of system clearly in terms of delay and interference level, the simulation results obtained by the proposed priority scheme are compared with RF1 and SFR. It is obvious that the proposed priority scheme can reduce the delay because the delay is one factor of

Table 4. System performance comparison for different coefficient values

W_1	W_2	W_3	Cell Edge Throughput [Kbps]	Mean Cell Throughput [Mbps]	Delay [ms]	Interference [dB]
0.33	0.33	0.33	233.92	8.64	10.13	-89.06
0.2	0.6	0.2	230.84	8.48	11.35	-88.94
0.2	0.2	0.6	209.07	8.40	9.88	-89.08
0.6	0.2	0.2	274.55	9.14	11.24	-87.83

weighting algorithm (see Figure 5). When the HoL of a UE increases, its priority increases and then it will be scheduled to transmit data with higher Modulation and Coding Scheme. Moreover, Figure 6 demonstrates that the interference level of the proposed priority scheme is lower than RF1 and SFR because the transmit power of each RB is calculated dynamically by exchanging messages among neighbouring eNBs. As a result, the interference level on each RB in each cell decreases.

The Figure 7 depicted that the proposed priority scheme can significantly improve the 5% of UE's throughput than RF1 and SFR due to considering interference level. In addition, Figure 8 demonstrates that the proposed priority scheme has better cell edge user throughput than RF1 and SFR schemes. Simulation result shows that the cell edge throughout of the proposed priority scheme is 68% higher than SFR and 96% higher than RF1.

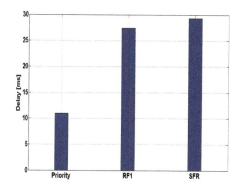

Figure 5. Delay comparison among priority, RF1 and SFR schemes

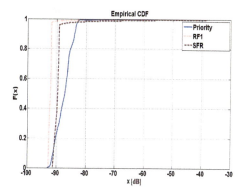

Figure 6. Interference level comparison among priority, RF1 and SFR schemes

In the proposed priority scheme, real time coordination is performed among all involved cells to avoid the use of same RB in two CEU in neighbouring cells. This is because the interference level of UEs has been considered as a parameter to find the priority. Moreover, unlike the RF1 and SFR schemes, the transmit power allocated to each UE is dynamically calculated through a FLS. The dynamic power allocation can lead to a reduction of interference for UEs particularly for CEU. Therefore, SINR increases and then a higher CQI value is selected. On the other side, CEUs and CCUs can be scheduled within the available bandwidth not only in a specified subband such as SFR. It leads to increase in the number of RBs which can be allocated to each UE. In RF1 and SFR the RB and power allocation in each cell is pre-defined and not adapted to the changes in network traffic.

Figure 9 compares the average cell throughput of the proposed priority scheme with RF1 and SFR. It shows that the proposed priority scheme has higher average system performance than SFR and RF1 around 12% and 28.7% respectively. This is because it dynamically adjusts the transmit power of each RB based on SINR, throughput and achievable data rate through the FLS. Moreover, it takes into account the system traffic loads as well as system changes when the RBs are allocated to different UEs.

6. CONCLUSION

A priority based ICIC scheme has been proposed in this paper to mitigate the ICI problem in LTE downlink system. The proposed priority scheme determines the priority of each UE using interference level, Quality of Service and Head of Line. UEs could be scheduled based on their priority on the specified subbands. Therefore, the network changes and traffic model can be considered in this method. The adaptive transmit power is determined through a fuzzy logic system. Simulation results showed that the proposed priority scheme outperforms the RF1 and SFR schemes in terms of cell edge user throughput and system throughput. As the future work, we will propose an enhanced intercell interference coordination algorithm to mitigate interference in LTE-Advanced heterogeneous networks.

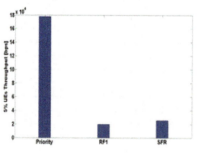

Figure 7. 5 % UE throughput comparison

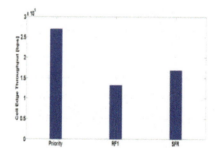

Figure 8. Average cell edge user throughput comparison

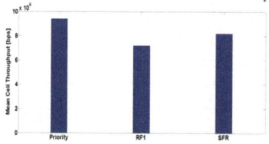

Figure 9. Average cell throughput comparison

REFERENCES

[1] I.F. Akyildiz, D. M. G. Estevez, E.Ch.Reyes, (2010) "The evolution to 4G cellular systems: LTE-Advanced", Physical Communication, Vol.3, No.4, pp. 217-244.

[2] F. Khan, (2009) . LTE for 4G Mobile Broadband - overview of Inter-Cell Interference Coordination in LTE, United States of America by Cambridge University Press.

[3] C. Gessner, (2008).UMTS Long Term Evolution (LTE) Technology Introduction,Rohde & Schwarz Products.

[4] D. Astely, E. , Dahlman, A. Furuskar, Y. Jading, M. Lindstrom, and S. Parkvall, (2009) "LTE: the evolution of mobile broadband", IEEE Communications Magazine, Vol.47, No.4, pp.44-51.

[5] E. Lawrey, (1999) "multiuser OFDM", IEEE International Symposium on signals Processing and its Applications, Vol.2, pp. 761-764.

[6] F.Khan, (2009).LTE for 4G Mobile Broadband: Air Interace Technologies and performance, Cambridge University Press.

[7] X. Zhang, Ch.He, L.Jiang, and J.Xu , (2008) "Inter-cell interference coordination based on softer frequency reuse in OFDMA cellular systems", International Conference on Neural Networks and Signal Processing, pp.270 – 275.

[8] Siemens, R1-060135, (2006) "Interference Mitigation by Partial Frequency Reuse", 3GPP RAN WG1#42, London, UK.

[9] Huawei, R1-050507, (2005) "Soft Frequency Reuse Scheme for UTRANLTE", 3GPP RAN WG1#41, Athens, Greece.

[10] Huawei, (2005) "Further Analysis of Soft Frequency Reuse Scheme".

[11] Huawei, (2005) "Soft Frequency Reuse Scheme for UTRAN LTE".

[12] Ericsson, R1-061374, (2006)"Downlink intercell interference coordination/avoidance evaluation of frequency reuse", 3GPP Project Document.

[13] R. Ghaffar and R. Knopp, (2010) "Fractional frequency reuse and interference suppression for OFDMA networks", 8th International Modelling and Optimization in Mobile, Ad Hoc and Wireless, pp. 273 – 277.

[14] F.B. Mugdim, (2007)"Interference Avoidance Concepts," WINNER II project.

[15] E. Haro, S. Ruiz, D. Gonzalez, M.G-Lozano, and J. Olmos, (2009).Comparison of Different Distributed Scheduling Strategies for Static/Dynamic LTE Scenarios, Technical University of Wien.

[16] IEEE 802.16 Broadband, (2008) "Fractional Frequency Reuse in Uplink", LG Electronics.

[17] M. Porjazoski and B. Popovski, (2010) "Analysis of Intercell interference coordination by Fractional frequency reuse in LTE", in Proc. Int Software, Telecommunications and Computer Networks (SoftCOM) Conference, pp. 160-164.

[18] M. Porjazoski and B. Popovski, (2010) "Contribution to analysis of Intercell interference coordination in LTE: A fractional frequency reuse case", Global Mobile Congress (GMC), pp.1-4.

[19] A. Mills, D. Lister and M. DeVos, (2011) "Understanding Static Inter-cell Interference Coordination Mechanisms in LTE", Journal of Communications, Vol.6, No.4, pp. 312-318.

[20] S-E. Elayoubi, O. BenHaddada, and B. Fouresti´e, (2008) "Performance Evaluation of Frequency Planning Schemes in OFDMA based Networks", IEEE Transaction on Wireless Communications, Vol.7, No.5, pp.1623-1633.

[21] Y.Yu, E. Dutkiewicz, X. Huang, and M. Mueck, (2012) " A Resource Allocation Scheme for Balanced Performance Improvement in LTE Networks with Intercell Interference" IEEE Wireless Communication and Networking conference: MAC and Cross-layer Design, pp. 1630-1635.

[22] L. Liu, G. Zhu, D. Wu, (2011) "A Novel Fractional Frequency Reuse Structure Based on Interference Avoidance Scheme in Multi-cell LTE Networks", 6th International ICST Conference on Communications and Networking in China (CHINACOM), pp.551 – 555.

[23] L. Liu, D. Qu, and T. Jiang, (2010) "Dynamic Fractional Frequency Reuse based on Interference Avoidance Request for Downlink OFDMA Cellular Networks", IWCMC, Caen, France, pp.381-386.

[24] L. Dong, and W. Wenbo, (2009) "A Novel Semi-Dynamic Inter-Cell Interference Coordination Scheme Based on UE Grouping", IEEE 70th Vehicular Technology Conference, pp.1-5.

[25] Ch.You, Ch.Seo, Sh. Portugal, G. Park, T. Jung, H. Liu and I. Hwang, (2011) "Intercell Interference Coordination Using Threshold-Based Region Decisions," Wireless Personal Communications, Vol. 59, No.4, pp.789-806.

[26] L. Shuhui, C. Yongyu, Y. Ruiming, and Y. Dacheng, (2011) "Efficient Distributed Dynamic Resource Allocation for LTE Systems", Vehicular Tecnology conference (VTC), pp.1-5.
[27] M. Iturralde, A. Wei, T-Ali. Yahiya, and A-L. Beylot, (2012) "Resource allocation for real time services using cooperative game theory and a virtual token mechanism in LTE networks", IEEE Conference on Consumer Communications and networking (CCNC), pp.879-883.
[28] M. Rahman, and, H. Yanikomeroglu, (2010) "Enhancing cell-edge performance: a downlink dynamic interference avoidance scheme with intercell coordination", IEEE Transaction on Wireless Communications, Vol.9, No.4, pp.1414-1425.
[29] M. Rahman, H. Yanikomeroglu, and W. Wong, (2009) "Interference Avoidance with Dynamic Inter-Cell Coordination for Downlink LTE System", IEEE Wireless Communications and Networking Conference (WCNC), pp.1-6.
[30] H.W. Khun, (1995) "The Hungarian method for the assignment problem," Naval Research Logistic Quarterly, Vol. 2, pp. 83-97,
[31] P. Vlacheas, E. Thomatos, K. Tsagkaris, and P. Demestichas, (2011)"Autonomic downlink inter-cell interference coordination in LTE Self-Organizing Networks", 7th International Conference on Network and Service Management (CNSM), pp.1-5.
[32] M.C. Necker, (2009) "Scheduling Constraints and Interference Graph Properties for Graph-based Interference Coordination in Cellular OFDMA Networks", Journal of Mobile Networks and Applications, Vol.14, No.4, pp.539-550.
[33] G.I.Fraimis , V. D.Papoutsis, and S.A. Kotsopoulos, (2010) "A distributed radio resource allocation algorithm with interference coordination for multi-cell OFDMA systems", IEEE 21st International Symposium on Personal Indoor and Mobile Radio Communications (PIMRC), pp. 1354 - 1359.
[34] Q. Ai, P. Wang, F. Liu, Y. Wang, F. Yang and J. Xu, (2010) "QoS-guaranteed cross-layer resource allocation algorithm for multiclass services in downlink LTE system", International Conference on Wireless Communications and Signal Processing (WCSP), pp.1-4.
[35] Z. Bingbing, C. Liquan, Y. Xiaohui, and W. Lingling, (2010) "A Modified Inter-Cell Interference Coordination Algorithm in Downlink of TD-LTE", 6th International Conference on Wireless Communications Networking and Mobile Computing (WiCOM), pp.1-4.
[36] Kh. Elsayed, (2011) "4G++: Advanced Performance Boosting Techniques in 4th Generation Wireless Systems and Beyond", The First NTRA Knowledge Dissemination and Networking Conference, Cairo, Egypt.
[37] G. Boudreau, J. Panicker, N. Guo, R. Chang, N. Wang, and S. Vrzic, Nortel, (2009) "Interference Coordination and Cancellation for 4G Networks", IEEE Communications Magazine, Vol. 47, No. 4, pp.74 – 81.
[38] R. Kwan and C. Leung, (2010) "A Survey of Scheduling and Interference Mitigation in LTE," Journal of Electrical and Computer Engineering, Vol.2010, pp.1-10.
[39] N. Himayat, Sh. Talwar, A. Rao and R. Soni, (2010) "Interference management for 4G cellular standards", IEEE Communications Magazine,Vol.48, No.8, pp.86-92.
[40] E. Pateromichelakis, M. Shariat, A. Quddus, and R.Tafazolli, (2012) " On the Evalution of multi-cell Scheduling in 3GPP LTE/LTE-A", IEEE Communication Surveys and Tutorials, issue.99, pp.1-17.
[41] 3GPP, TS 23.203, (2012) "Digital cellular telecommunications system (Phase 2+); Universal Mobile Telecommunications System (UMTS); LTE; Policy and charging control architecture", V10.6.0.
[42] L. Zadeh, (1965)"Fuzzy sets", Information Control, Vol. 8,No.3, pp.338–353.
[43] T. Ross, (1995). Fuzzy Logic with engineering application, McGraw-Hill.
[44] S. Sesia, I. Toufik, M. Baker, (2011). Lte - The Umts Long Term Evolution: From Theory to Practice, Wiley, pp. 1-648, ISBN: 0470697164, 9780470697160.
[45] G. Roche, A. Al-Glazunov, and B. Allen, (2012). LTE-Advanced and Next Generation Wireless Networks: Channel Modelling and Propagation, Willey, pp.1-566, ISBN: 978-1-1199-7670-7.

Authors

Ameneh Daeinabi is a PhD student in Telecommunications Engineering at University of Technology Sydney, Australia and her current research area is 4G mobile networks. She received her M.Sc. degree in Telecommunications Engineering from Sahand University of Technology in 2009 and her B.Sc. in Electrical &Electronic Engineering from Azad University, in 2005. Her main research interests include 4G mobile networks and Vehicular Ad hoc Networks.

Kumbesan Sandrasegaran is an Associate Professor at UTS and Director of the Centre for Real-Time Information Networks (CRIN). He holds a PhD in Electrical Engineering from McGill University (Canada)(1994), a Master of Science Degree in Telecommunication Engineering from Essex University (1988) and a Bachelor of Science (Honours) Degree in Electrical Engineering (First Class) (1985). He was a recipient of the Canadian Commonwealth Fellowship (1990-1994) and British Council Scholarship (1987-1988). His current research work focuses on two main areas (a) radio resource management in mobile networks, (b) engineering of remote monitoring systems for novel applications with industry through the use of embedded systems, sensors and communications systems. He has published over 100 refereed publications and 20 consultancy reports spanning telecommunication and computing systems.

Xinning Zhu is an associate professor at School of Information and Communication Engineering, BUPT, China. She received her PhD, MS and BS degree in Communication and Information system from Beijing University of Posts and Telecommunications (BUPT) in 2010, 1995 and 1992. Her current research interests focus on interference management and mobility management in radio resource management for heterogeneous networks.

RELIABILITY OF MOBILE AGENTS FOR RELIABLE SERVICE DISCOVERY PROTOCOL IN MANET

[1]Roshni Neogy   [2]Chandreyee Chowdhury   [2]Sarmistha Neogy*

[1] Dept. of Information Technology, Jadavpur University
[2]Dept. of Computer Science & Engineering, Jadavpur University
*sarmisthaneogy@gmail.com

ABSTRACT

Recently mobile agents are used to discover services in mobile ad-hoc network (MANET) where agents travel through the network, collecting and sometimes spreading the dynamically changing service information. But it is important to investigate how reliable the agents are for this application as the dependability issues(reliability and availability) of MANET are highly affected by its dynamic nature.The complexity of underlying MANET makes it hard to obtain the route reliability of the mobile agent systems (MAS); instead we estimate it using Monte Carlo simulation. Thus an algorithm for estimating the task route reliability of MAS (deployed for discovering services) is proposed, that takes into account the effect of node mobility in MANET. That mobility pattern of the nodes affects the MAS performance is also shown by considering different mobility models. Multipath propagation effect of radio signal is considered to decide link existence. Transient link errors are also considered. Finally we propose a metric to calculate the reliability of service discovery protocol and see how MAS performance affects the protocol reliability. The experimental results show the robustness of the proposed algorithm. Here the optimum value of network bandwidth (needed to support the agents) is calculated for our application. However the reliability of MAS is highly dependent on link failure probability.

KEYWORDS

Reliability, Mobile agents, Monte Carlo simulation, Mobile ad-hoc network, service discovery, Mobility Model

1. INTRODUCTION

A mobile agent is a combination of software program and data which migrates from a site to another site to perform tasks assigned by a user according to a static or dynamic route [1]. It can be viewed as a distributed abstraction layer that provides the concepts and mechanisms for mobility and communication [2]. An agent consists of three components: the program which implements it, the execution state of the program and the data. An agent may migrate in two ways, namely, weak migration and strong migration [3]. The platform is the environment of execution. The platform makes it possible to create mobile agents; it offers the necessary elements required by them to perform their tasks such as execution, migration towards other platforms and so on.

Typical benefits of using mobile agents include [4]
- Bandwidth conservation: sending a complex query to the database server for processing.
- Reduced latency: a lightweight server can move closer to its clients
- Load balancing: loads may move from one machine to the other within a network etc.

The route of the mobile agent can be decided by its owner or it can decide its next hop destination on the fly. Here, we assume the underlying network to be a Mobile Ad Hoc Network (MANET) that typically undergoes constant topology changes, which disrupt the flow of information over the existing paths. Mobile agents are nowadays used in MANETs for various purposes like service discovery [5], network discovery, automatic network reconfiguration etc. But before mobile agent based applications become commercially available for MANET, reliability estimation of them is very essential. Because of motion and location independence [1], this environment itself introduces new aspects to reliability (in terms of continuity of correct service).

In [1] [6] we tried to address this issue. In [7] reliability estimation of service discovery [5] agents is considered along with few modifications to the basic mechanism [5] in order to reflect the dynamism of the underlying environment. But node mobility is not considered explicitly, the link connectivity is assumed to follow Non Homogenous Poisson (NHPP) distribution. Thus we did not consider parameters like pattern of node mobility, transient failure of the links etc.

In this paper node mobility is represented by mobility models. Three kinds of mobility are considered. Random movement of nodes is proposed in Random Waypoint Mobility Model [8](RWMM), node movement based on temporal dependency in Smooth Random Mobility Model [9] (SRMM) and node movement based on spatial dependency in Reference Point Group Mobility Model [10] (RPGM). Also multipath propagation of radio signals is considered according to two ray propagation model [11] as it is simple and widely used in literature. In reality, even when an agent finds a node to be connected to where it is now residing, an attempt to migrate to that node (host site) may fail because of transient link errors like frequency selective fading, heavy rainfall etc. This effect can be modelled as a Poisson process. Reliability estimation algorithm of [7] is modified to incorporate these changes. Also a metric is proposed to measure the performance of the service discovery protocol. This enables us to thoroughly study the effect of MAS reliability on service discovery. Our MAS reliability estimation model is not tightly coupled to the application (service discovery) but with minor modifications it can be applied to agents deployed by other MANET applications as well.

In the following section we discuss about the service discovery process using mobile agents in MANET. Then in section 3 state of art regarding this topic is mentioned. In section 4 our model is introduced that is designed to estimate reliability of the mobile agent based system (MAS). The next section (5) gives the experimental results followed by concluding remarks in section 6.

2. THE PROCESS OF SERVICE DISCOVERY

A service can be regarded as any hard- or software resource, which can be used by other clients. Service discovery is the process of locating services in a network. The following methods are used to discover and maintain service data [12]:
- service providers flood the network with service advertisements;
- clients flood the network with discovery messages;

- nodes cache the service advertisements;
- nodes overhear in the network traffic and cache the interesting data.

The first one corresponds to passive discovery (push model) whereas the next one describes active discovery (pull model). The other two methods mentioned above are the consequences of the first two. While the push mechanism is quite expensive in terms of network bandwidth (in the context of MANET), the pull mechanism suffers from poor performance (longer response times). Moreover there are other factors to be taken into account such as the size of the network (no. of nodes), availability of a service (how frequently services appear and disappear in the network), and the rate of service requests. Traditionally static service brokers are used for sharing service information which is not suitable for MANET due to its inherent dynamic nature. So as in [5] mobile agents can be deployed for this purpose (looking for services offered) as the agents can migrate independently [13], behave intelligently [14] and can negotiate with other agents according to a well defined asynchronous protocol [15].

The service discovery protocol presented in [5] is taken to be the basis here. We first estimate the reliability of MAS where the agents are roaming around the underlying MANET, discovering various services provided by the nodes in MANET. To do this the algorithm [5] uses two types of agents – a static Stationery Agent (SA) and mobile Travel Agent (TA). The SAs are deployed on per node basis. On the contrary the TAs are deployed dynamically to collect and spread service information among the nodes in MANET. A TA prefers those nodes on its route which it has not yet visited but which are reachable via nodes it already knows. In order to enforce this TA Route algorithm is proposed in [5] that determines the next target migration site of a TA. The SAs are responsible for controlling the no. of TAs roaming around the network. Thus depending on the incoming agent frequency (that is, no. of agents TA visiting a node is said to be incoming agent frequency) of TA, an SA can either create or terminate a TA depending on network bandwidth. Larger the bandwidth more agents can be supported leading to better performance and probably improved reliability.

3. RELATED WORKS

Reliability analysis of MAS in ad-hoc network is a complicated problem for which little attention has been paid. Most of the work done in this area is related to distributed systems and distributed applications. But as pointed out in [16], features like scalability and reliability becomes critical in challenging environment with wireless networks. However the scalability/reliability issue of MAS has been highlighted in [17] although the work does not focus on resource constrained environments like MANET. Moreover this work does not take into account the specific task for which the agents are deployed. But this is very much important as route of a mobile agent primarily depends on the purpose for which it is deployed. However, we could not find any work that considers estimation of reliability of service discovery agents for MANET but we found the following.

3.1. Service Discovery in MANET

There are already some approaches for service discovery in MANETs. Some of them attempt to optimize flooding by reducing its overhead [18], but they stillcause a lot of traffic. In [12] some device and service discovery protocols are discussed along with the issues in MANET. However the work does not provide a detailed concrete solution to the problem of service discovery though it suggests possible use of mobile agents in discovering services. In [19] an overlay structure is used to distribute service information. Here the network is divided into groups of nodes and nodes share service information among the group members. Only if a service request is not addressed in the present group, then it will be forwarded to the adjacent group. But in highly dynamic scenario this group formation can become an overhead. To reduce such overhead in [5] mobile agents are used. But this work does not take into consideration the movement of nodes before an agent finishes its job. Moreover this algorithm expects the network to retain the same connectivity pattern while an agent is roaming around the MANET.

3.2. Reliability in MANET

Due to the analytical complexity and computational cost of developing a closed-form solution, simulation methods, specifically Monte Carlo (MC) simulation are often used to analyze network reliability. In [20], an approach based on MC method is used to solve network reliability problems. In this case graph evolution models are used to increase the accuracy of the resultant approximation.

But little has been addressed on the reliability estimation of MANETs. In [21] analytical and MC-based methods are presented to determine the two-terminal reliability for the adhoc scenario. Here the existence of links was considered in a probabilistic manner to account for the unique features of the MANET. However, there remains a gap in understanding the relationship between a probability and a specific mobility profile for a node. In [22] MC-based methods are presented to determine the two-terminal reliability for the adhoc scenario. This work is an extension of that in [21] by including directly, mobility models in order to allow mobility parameters, such as maximum velocity, to be varied and therefore analyzed directly. The methods in this paper will now allow for the determination of impacts of reliability under specific mobility considerations. As an example, one may consider the different reliability estimate when the same networking radios are used to create a network on two different types of vehicles. Here node mobility is simulated using Random Waypoint mobility model [8]. But this Random Waypoint model of mobility being a very simple one often results in unrealistic conclusions. Moreover none of this work focuses on mobile agents but only the 2-terminal [8] or all terminal network reliability.

3.3. Reliability of Mobile Agents

Little attention has been given to the reliability analysis of MAS. In [23], two algorithms have been proposed for estimating the task route reliability of MAS depending on the conditions of the underlying computer network. In [24], which is an extension of the previous work, a third algorithm based on random walk generation is proposed. It is used for developing a random static planning strategy for mobile agents. However, in both the works the agents are assumed to be independent and the planning strategy seemed to be static. So this work does not address the scenario where agents can change their routes dynamically. Moreover, it does not address the issue of node mobility in between agent migrations.

In [1] a preliminary work has been done on estimating reliability of independent mobile agents roaming around the nodes of a MANET. The protocol considers independent agents only. Node and link failure due to mobility or other factors is predicted according to NHPP. An agent is allowed to migrate to any node with equal probability. This may not be realistic as some nodes may provide richer information for a particular agent deployed by some application. In [6] the MAS is assumed to be consisting of a number of agent groups demanding a minimum link capacity. Thus, each agent group requires different channel capacity. Hence, different groups perceive different views of the network. In this scenario the reliability calculation shows that even with large no. of heterogeneous agent groups with differing demands of link capacity, the reliability of the MAS gradually reaches a steady state. Since the task (for example service discovery) given to an agent primarily controls its routes, it is an important aspect and must be considered while estimating reliability. But the nature of the task and hence the mobile agent's movement pattern is not considered in any of these works.

4. OUR MODEL

Though mobile agents (MA) are recently used in many applications of MANET including service discovery, dependability analysis of such applications is not much explored. In [7] reliability analysis of service discovery agents is attempted. Here the agents will tend to migrate towards the crowded portion of MANET to collect and fast spread service information. But the work does not consider many insights like the effect of node mobility or transient faults. Also performance of service discovery protocol with respect to agent's performance is not considered. In the present workthat is an extended version of [7], the effect of underlying environment on service discovery agents are considered. Our model is described in three parts - modelling of MANET, modelling of service discovery agents on MANET and reliability estimation of mobile agent system

4.1. Modelling MANET

We model the underlying network as an undirected graph G= (V,E) where V is the set of mobile nodes and E is the set of edges among them. Let the network consist of N nodes, thus |V|=N that may or may not be connected via bidirectional links. The following assumptions are made ([25][26]):

- The network graph has no parallel (or redundant) links or nodes.
- The network graph has bi-directional links.
- There are no self-loops or edges of the type (v_j, v_j).
- The states of vertices and links are mutually statistically independent and can only take one of the two states: working or failed.

Modeling node mobility is also one of the big problems in MANET. Many mobility models are proposed that can address this issue based on a specific application scenario such as campus network or urban areas. RWMM [8] is widely used in this regard for its simplicity. Here a mobile node randomly chooses a destination point (waypoint) in the area and moves with constant speed on a straight line to this point. After waiting a certain pause time, it chooses a new destination and speed, moves with constant speed to this destination, and so on. Here we have chosen a linear velocity $v_i(t)$ and a direction $\varphi_i(t)$ and calculate the next destination point as

$$x_i(t+\Delta t) = x_i(t) + \Delta t * v_i(t) * \cos\varphi_i(t) \quad (1)$$

$$y_i(t+\Delta t) = y_i(t) + \Delta t * v_i(t) * \sin\varphi_i(t) \quad (2)$$

But Random Waypoint can result in a sharp turn or sudden stop when the differential rate of change of velocity is infinity which is not feasible in a practical scenario [9]. So to smoothen such change in velocity, and hence make the model more realistic, SRMM [9] model is also considered. It can be simulated as in [1]

$$x_i(t+\Delta t)=x_i(t)+\Delta t*v_i(t)*\cos\varphi_i(t)+0.5*a_i(t)*\cos\varphi_i(t)*\Delta t^2 \qquad (3)$$

$$y_i(t+\Delta t)=y_i(t)+\Delta t*v_i(t)*\sin\varphi_i(t)+0.5*a_i(t)*\sin\varphi_i(t)*\Delta t^2 \qquad (4)$$

In situations like battlefield or rescue work people work in groups and hence movement in a group is commonly observed in such situations. So we can also use RPGM[10] in our simulation. In that case the speed and direction (angle) of mobile nodes (MNs) would follow that of their leader, called a reference point. The velocity of the leader can follow RWMM again. If v_{leader} and φ_{leader} represent the speed and direction of movement of the reference point respectively then the speed (v_i) and direction (φ_i) of MN_i can be calculated as follows [10]

$$v_i(t) := v_{leader}(t) + random()*SDR*v_{max} \qquad (5)$$

$$\varphi_i(t) = \varphi_{leader}(t) + random()*ADR*\varphi_{max} \qquad (6)$$

Here SDR and ADR are speed and angle deviation ratio respectively having the following relation 0<SDR, ADR<1. Hence position of MN_i at ($t+\Delta t$) time instant can be estimated as

$$x_i(t+\Delta t) = x_i(t) + \Delta t*v_i(t)*\cos\varphi_i(t) \qquad (7)$$

$$y_i(t+\Delta t) = y_i(t) + \Delta t*v_i(t)*\sin\varphi_i(t) \qquad (8)$$

Thus the movement of the nodes are simulated in one of the above three ways. Now the received signal power is calculated according to two ray propagation [11] of radio signals to handle multipath propagation effect. It is supplemented in Shannon's theorem to calculate link capacity as in [6]. A failed link can be treated as a link having zero capacity. Link capacity also varies with time even if the adjoining nodes do not change their position thus featuring transient nature of the faults.

Thus an initial configuration would be assumed. Afterwards due to mobility few links may fail and still a few may be revived also according to NHPP considering the transient nature of the faults.

4.2. Modelling Service Discovery Agents on MANET

In this paper, we assume that our MAS (S) at a time instant has m(t) independent agents (Travel Agents in [5]) that may move in the underlying MANET. Here m(t) indicates the fact that the no. of TAs varies with time as an SA can kill TAs [5]. The reliability of (S) is defined as the probability that (S) is operational during a period of time [2]. Later we define reliability of an individual agent in this context. The commonly used terms are listed in Table 1.

In this scenario we can think of an agent as a token visiting one node to another in the network (if the nodes are connected) based on the strategy listed as TA Route Algorithm in [5]. But node mobility in between agents' journey was not considered in [5]. So we have made necessary modifications to make the service discovery process more suited to the dynamics of MANET. A TA starts its journey from an owner (where it is created by SA) and moves from one node to another according to the TA route Algorithm [5]. But this movement is successful if the two nodes are connected and there is no simultaneous transmission in the neighbourhood of the intended destination (taken care of by the MAC protocol). So, we associate a probability with the movement to indicate transient characteristics of the environment, since, for example, the routing table may not be updated properly or the link quality may have degraded so much (due to increased noise level) that the agents are unable to migrate. Thus, if an agent residing at node MN_A decides to move to node MN_B (connected to MN_A) then the agent successfully moves to MN_B with probability p_t. Here p_t denotes the problem of unpredictability mentioned above. For example, noise level may increase due to heavy rainfall. If at any time an agent finds all unvisited nodes to be unreachable, the agent waits and then retries. This step tolerates the transient faults (temporary link failure) as an agent retries after some delay and hence improves system performance. This is not considered in [5] but to make the service discovery process more suitable to the MANET dynamicity, transient fault tolerance becomes a necessity.

Table 1. Notation

Terms	Descritption
$m(t)$	No. of Agents in the system at time t
$\lambda_i(t)$	task route reliability of ith agent in a step of simulation
$\lambda(t)$	average reliability of all agents
$r_i(t)$	probability that m_i is working correctly at time t, that is, the individual software reliability of mi
SP_{total}	total no. of service providers
SP_{dis}^i	no. of service providers discovered by i_{th} node
$SP_{collected}$	No. of service providers discovered by an agent
$R_i(t)$	instantaneous reliability of service discovery
$R_{service}^i$	reliability of service discovery protocol till time T
Q	no. of simulation steps

In this scenario we study the reliability of MAS (consisting of the TAs) with respect to the network status and its conditions (for example connectivity of the links, path loss probability etc.). Each agent is expected to visit all operating nodes in MANET in order to collect and spread service information. We have taken the failure probability (P) of the mobile nodes (P_{Node}) to be a variable of Weibull distribution [6].

Now reliability of MAS (R_s) can be defined as

$$R_s = \{R_{MAS}|R_{MANET}\} \quad (9)$$

This is a conditional probability expression indicating the dependence of MAS reliability on conditions of MANET. Here reliability of MANET (R_{MANET}) can be treated as an accumulative factor of $(1-P_{Node})$ and P_{Link}. P_{Link} can be treated as a combination of P (p_r is at an acceptable level) and the mobility model. Here p_r denotes the received power at node j after traversing distance d_{ij} from sender node i. Here we calculate individual agent reliability on the underlying MANET as follows:

If an agent can successfully visit M nodes out of N(desired) then it has accomplished M/N portion of its task. Thus reliability in this case will be M/N.

But if the application requires all N nodes to be visited to complete the task and in all other cases the task will not be considered to be done, the calculation will be modified as:

If an agent can successfully visit all N nodes desired then it has accomplished its task. Thus reliability in this case will be 1. In all other cases it will be 0.

Above definitions of agent reliability works only if there is no software failure of the agent (assumed to follow Weibull distribution [6]).

Now, the probability that the MAS is operational i.e., reliability of MAS (R_{MAS}) can be calculated as the mean of reliability of all its components, that is, the agents in this system. Clearly it is function (m(t)) of time as the total no. of TAs present varies with time

$$R_{MAS}(t) = \frac{\sum \{AgentReliabilities\}}{m(t)} \quad (10)$$

However the performance of the service discovery protocol is measured by a node's success in discovering service providers in the network. All or a few nodes in MANET may act as service providers, thus $SP_{total} \leq N$. The instantaneous reliability of MN_i ($R_i(t)$) can be given as

$$R_i(t) = \frac{SP_d^i}{SP_{to}} \quad (11)$$

If $R_i(t)$ is integrated over time, it gives the overall service discovery reliability $R_{service}^i$ as follows

$$R_{service}^i = \frac{1}{T}\int_0^t R_i(t)dt = \frac{1}{T}\sum_T R \quad (12)$$

Here time is divided in discrete time steps and snapshot of the system is taken in short periodic intervals. As this interval tends to 0 a near continuous picture of S can be observed. The same is repeated Q times according to Monte Carlo simulation [1]. The following algorithm presented in this paper describes the reliability analysis of agents deployed for discovery and spreading of service information among the nodes in MANET. Reliability of the agent system is calculated according to equation 10 while equation 11 along with 12 calculates the performance of individual nodes.

4.3. Detailed steps of reliability estimation

1) Input parameters: M (initial no. of TAs in the system), the initial state of the network (node position, location, speed of the nodes)
2) Detailed Steps:
   The TA determines its next target according to the following algorithm.
   TA_Route_Mod()
1. First the travel agent tries to find yet unvisited nodes, which are common neighbors of previously visited nodes. These common neighbor nodes have highest priority because their services can be used directly by more than one node.
2. If all such common neighbors have been visited then the agents will next visit the nodes not visited yet. If there are more than one unvisited neighbors, the mobile agent can choose to visit any one of them.

3. If there are no unvisited nodes in direct range of a mobile agent then a node with unvisited neighbors is revisited. If there are two such potential nodes, the node with the lowest RSN is chosen.
4. If the first three conditions fail, the node with the lowest RSN becomes the node visited next.

Node mobility is considered here and is detailed in the following algorithm. As such the agent's decision in choosing the next destination depends on the currently reachable set of nodes. Here we are hoping that at short time intervals the network topology will not show huge changes so as to render the common neighboring nodes absolutely disconnected or be reduced to a node with very few neighbors (less than the number of nodes which influence the MA's decision to grant it the highest priority).

Reliability_Calculation()
1. Initialize n (that is the no. of mobile nodes successfully visited by an agent) to 0 and a source for the mobile agent.
2. Input network configuration (V, E) in the form of an edge list.
3. Node mobility is simulated according to RWMM/SRMM/RPGM.
    3.1. Node connectivity received signal power (P_r) (according to Two ray propagation model) and hence node connectivity is calculated as in [14].
    3.2. Some nodes may also fail because of software/hardware failure or become disconnected from the network according to another NHPP (or uniform) distribution. Node failure can be simulated by deleting the edges e from E' further that are incident on the failed node.
4. According to Weibull distribution we find individual software reliability r_i for an agent i.
5. Breadth First Search (BFS) is used unless all connected subgraphs are assigned a proper cluster id. Thus, an isolated node is also a cluster.
6. The agents perform their job on this modified graph according to *TA_Route_Mod()*.
7. Repeat step 6 for all agents (m(t)) in the system.
8. When an agent comes back its $SP_{collected}$ is used to find its reliability, $\lambda_i(t)$ according to equation 11 and 12.
    8.1. Spawn new agents if service information is still incomplete.
9. Repeat steps 3 to 7 until all nodes are visited or the new destination falls in a different cluster.

10. Calculate $$\lambda_i(t) = \frac{n}{N} \qquad (13)$$

Here the value of n depends heavily on the conditions of the underlying network.

11. Reset the value of n.

12. Calculate $$\lambda(t) = \frac{1}{k}\sum_{i=1}^{k} \lambda_i(t) r_i \qquad (14)$$

13. Repeat steps 3 to 12 Q (simulation steps) times.

14. Calculate node reliability $$\frac{1}{Q}\sum_{q=1}^{Q} \lambda(q,t) \qquad (15)$$

It is to be noted that step 3 is repeated for every move of the agent to take care of network dynamicity. If an agent fails to move because of background noise level, then it may retry depending on the amount of delay that the respective application can tolerate.

5. RESULTS

The simulation is carried out in java and can run in any platform. The initial positions of MAs and the initial network configuration are read from a file. The default values of parameters are listed in Table 2. Unless otherwise stated, the parameters always take these default values. Detailed analysis of simulation results is shown.

In reality network dynamicity affects agent migration and hence the reliability of MAS is found to depend heavily on its size (no. of agents) particularly for bigger MANETs. This fact is shown in figure 1. This graph is taken for two different scenarios. In one of them, each node can receive maximum (maximum tolerated frequency) 20 agents (gray columns), additional agents would get killed resulting in a drop in agent reliability. For the other one, the maximum tolerated frequency is kept at 15. For $m(t)$ ($<=15$) as far as the underlying MANET remains connected, all agents will be able to complete their job if there is no software failure in them as shown in figure. But if $m(t)$ is increased any further, then agent reliability drops as it exceeds the maximum tolerated agent frequency. Higher the no. of agents tolerated in the network greater will be the overall agent reliability for bigger MAS. This justifies the right side of the graph in figure 1.

Table 2. Default values for simulation parameters

Parameter Name	Value
Number of nodes(N)	25
Number of agents(m(t))	25
Number of simulations(Q)	200
SP_{total}	2
Maximum Agent Frequency tolerated(MAX FR)	20
Link Failure Probability	0.2
Time for each simulation run	750min
Mobility Model	SRMM

Every MANET has a bandwidth limitation that in turn restricts the maximum value of $m(t)$ during a period. Thus, it can be observed that there is a maximum incoming agent frequency supported by a node (figure 2). Higher value of this indicates greater bandwidth provided by MANET. So as expected, with higher bandwidth ourMAS become more reliable. But it can be observed that with $m(t)=25$ when the incoming agent frequency reaches above 18, the MAS reaches an almost steady state with overall reliability of (around) 0.95. This gives the optimum value of bandwidth to be provided by MANET for this scenario.It is interesting to observe that for optimum performance all $m(t)$ agents need not be tolerated at a time (as $m(t)>$ maximum agent frequency tolerated).

Now if N is increased, the overall reliability reaches a steady state at N=20 as long as m(t) (<30) and hence is comparable to the agent frequency (=20) supported by the nodes (figure 3). Thus our approach is found to be scalable for MANETs as big as 45 nodes. But for large m(t) (>=30), nodes may kill appreciable number of agents resulting in a drop in reliability as a MANET grows. But this result indicates the scalability of the service discovery approach for crowded MANET as change in network size does not appreciably affect the reliability of MAS rather it helps the agents to discover service providers even faster by redundant routes.

The movement pattern of the nodes is found to play an important role in many MANET applications as it affects link existence probability. For smaller MANETs, nodes moving randomly according to RWMM exhibits better MAS reliability as compared to SRMM or RPGM (figure 4).

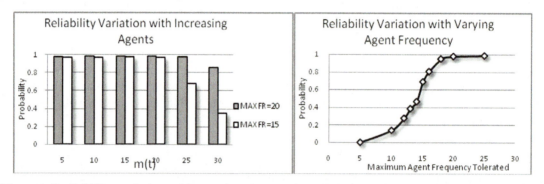

Figure1. Reliability variation with no.of agents   Figure2. Reliability variation with varying tolerance limit on accepted agent frequency

The reason can be periodicity in link existence probability [8]. But as a MANET grows, the performance of MAS becomes almost independent of the effect of different mobility models and hence user movement pattern. This result shows that the reliability prediction made in one scenario (say campus network following SRMM) is very much applicable in several other scenario (including disaster relief following RPGM).

Though mobility pattern of the nodes does not affect its agents but the transient failure of the links does. The link (transient) failure probability is found to affect reliability of the

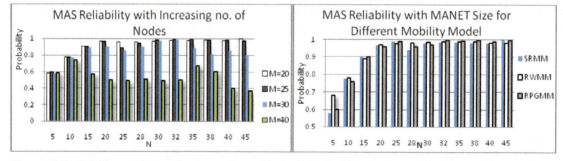

Figure 3.Reliability with varying no. of nodes(N)   Figure4. Reliability variation with MANET size for different mobility models

agents(figure 5). Thus even when the nodes are well within radio coverage of each other due to environmental effects like frequency selective fading or even heavy rainfall links may not be established. This is represented by the link failure probability (LFP), a metric that roughly follows NHPP. As more links fail, the network becomes partitioned resulting in a sharp fall in MAS reliability (figure 5). It can be observed that when the link failure probability reaches above 0.3, the network graph loses connectivity and as a result some agents may never reach the nodes residing in another component of the network.

The reliability of service discovery protocol is analyzed according to equations 11 and 12. The performance is measured from MN_{24}'s view. First the performance is measured with no. of service providers (SP_{total}). As SP_{total} increases service availability increases and hence reliability of the protocol after initial perturbation reaches a steady state (figure 6). But the effect of service availability is more apt in MANETs with more transient disturbance. As the

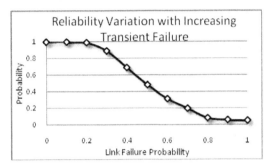

Figure 5. Reliability variation with varying link failure probability

environmental conditions stabilize the agents perform more reliably discovering almost all service providers even when they are scarce.

The effect of MANET size on the performance of service discovery is also studied as shown in figure 7. But as a MANET becomes more crowded (N>30) redundant paths emerge to reach service providers even when they are scarce. Here the total no. of service providers is taken to be 13. The drop in reliability for LFP=0.5 near 25 to 30 nodes MANET is because at this stage the nodes newly added increases the MANET boundary rather than making it more crowded. Thus the overall network connectivity worsens. But with stable environment (low transient errors) the service discovery protocol manages to discover most service providers.

6. CONCLUSION

In this paper, a scalable approach to estimate the reliability of MAS for MANET is presented. The agents are deployed for collecting and spreading service information in the network. The reliability is found to depend heavily on MANET dynamics in terms of transient link failure and supported network bandwidth. However the effect of different movement patterns of the users is found to affect MAS reliability a little. Thus conclusion drawn for one scenario (represented by a mobility model) may well be applied to others (other mobility models) if the transient failure probability <0.3. Moreover our approach is found to be scalable for bigger MANETs too. The agents choose their destination on route according to a service discovery algorithm based on [5]. During the time an agent is visiting a node, the underlying network may change according to SRMM taken to be the default case.

The protocol is validated and results are shown in section 5. As can be seen, reliability improves heavily if the network supports higher no. of agents. Hence as per expectation, this modified model works well in an efficient manner regardless of the pattern of changes in the network topology.A metric is proposed to measure reliability of the protocol itself and shows how agents help in discovering services overcoming the effect of poor network connectivity.

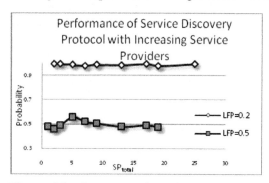 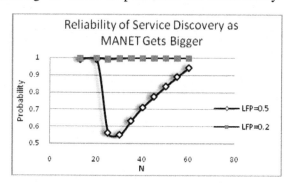

Figure 6.Performance of service discovery protocol with increasing service providers

Figure7. Performance of service discovery protocol with increasing N

The results of the simulation also corroborated our expectation. Our future work may include multiagent communication to make the process of service discovery even more reliable and efficient.We are also planning to include security issues in it.

REFERENCE

[1] C. Chowdhury, S. Neogy, "Estimating Reliability of Mobile Agent System for Mobile Ad hoc Networks", Proc. 3rd International Conference on Dependability, pp. 45-50, 2010.

[2] J. Cao, X. Feng , J. Lu , S. K. Das, "Mailbox-Based Scheme for Designing Mobile Agent Communications", Computer, v.35 n.9, pp 54-60, September 2002.

[3] N. Migas, W.J. Buchanan, K. McArtney, "Migration of mobile agents in ad-hoc, Wireless Networks", Proc.11th IEEE International Conference and Workshop on the Engineering of Computer-Based Systems, pp 530 – 535, 2004.

[4] D. B. Lang and M. Oshima, "Seven good reasons for mobile agents", in the Magazine Communications of the ACM, Vol. 42, Issue 3, March, 1999.

[5] R.T. Meier, J. Dunkel, Y. Kakuda, T. Ohta, "Mobile agents for service discovery in ad hoc networks", Proc. 22nd International Conference on Advanced Information Networking and Applications, pp 114-121, 2008.

[6] C. Chowdhury, S. Neogy, "Reliability Estimate of Mobile Agent Based System for QoS MANET Applications", in the Annual Reliability and Availability Symposium 2011, pp. 1-6, 2011.

[7] C. Chowdhury, S. Neogy, "Reliability Estimation of Mobile Agents for Service Discovery in MANET", in the International Conference on Parallel, Distributed Computing technologies and Applications (PDCTA 2011), pp. 148-157, 2011.

[8] C. Bettstetter and C. Wagner, "The Spatial Node Distribution of the Random Waypoint Mobility Model", in the Proceedings of German Workshop on Mobile Ad Hoc Networks (WMAN) (2002).

[9] C. Bettstetter, "Smooth is better than sharp: a random mobility model for simulation of wireless networks", in the Proceedings of the Fourth ACM International Workshop on Modeling, Analysis and Simulation of Wireless and Mobile Systems, pp. 19-25, 2001.

[10] X. Hong, M. Gerla, G. Pei, and C. Chiang. "A group mobility model for ad hoc wireless networks". In Proceedings of the ACM International Workshop on Modeling and Simulation of Wireless and Mobile Systems (MSWiM), August 1999.

[11] M. Rooryck, "Modelling multiple path propagation- Application to a two ray model", in the journal of L'Onde Electrique , ISSN 0030-2430, vol. 63, pp. 30-34, Aug.-Sept. 1983.

[12] J. Albert, S. Chaumette, "Device and Service Discovery in Mobile Ad-hoc Networks", Technical report, Master 2 SDRP, Universit´e Bordeaux 1, Jan. 16, 2007.

[13] M. Wooldridge, N. R. Jennings, "Intelligent agents - theory and practice", Knowledge Engineering Review 10 (2), pp. 115-152, 1995,.

[14] S. Ossowski, A. Omicini, "Coordination Knowledge Engineering", Knowledge Engineering Review 10 (2), pp. 115-152, 2002.

[15] J. Dunkel, R. Bruns: Software Architecture of Advisory Systems Using Agent and Semantic Web Technologies,Proceedings of the IEEE/ACM International Conference on Web Intelligence, Compiégne, France, IEEE Computer Society, pp. 418-421, 2005.

[16] O. Urra, S. Ilarri and E. Mena, "Agents jumping in the air:dream or reality", In 10th International Work-Conference on Artificial Neural Networks (IWANN'09), Special Session on Practical Applications of Agents and Multi-Agent Systems, pages 627–634. Springer, 2009.

[17] S. Ilarri, R. Trillo, E. Mena, "SPRINGS: A scalable platform for highly mobile agents in distributed computing environments", In: 4th International WoWMoM 2006 Workshop on Mobile Distributed Computing (MDC 2006), pp. 633–637. IEEE, Los Alamitos 2006.

[18] S. Helal, N. Desai, V. Verma, C. Lee: KONARK – A Service Dicovery and Delivery Protocol for Ad Hoc Networks. Proc. of the 3rd IEEE Conf. on Wireless Communication Networks (WCNC'03), Volume 3, pp. 2107-2113, 2003.

[19] M. Klein, B. Konig-Ries, and P. Obreiter. Lanes – A Lightweight Overlay for Service Discovery in Mobile Ad Hoc Networks. In Proc. of the 3rd IEEE Workshop on Applications and Services in Wireless Networks (ASWN2003), 2003.

[20] T.Elperin, I. Gertsbakh, M. Lomonosov, "Estimation of network reliability using graph evolution models", in the IEEE Transactions on Reliability, Vol. 40. No. 5. pp 572-581, 1991.

[21] J.L. Cook, J.E Ramirez-Marquez, "Two-terminal reliability analyses for a mobile ad-hoc wireless network", in Reliability Engineering and System Safety, Vol. 92, Issue 6, pp. 821-829, June 2007.

[22] J. L. Cook, J.E. Ramirez-Marquez, "Mobility and reliability modeling for a mobile ad-hoc network", IIE Transactions, 1545-8830, Vol. 41, Issue 1, pp. 23 – 31, 2009.

[23] M. Daoud, Q. H. Mahmoud, "Reliability estimation of mobile agent systems using the Monte Carlo approach", Proc. 19th IEEE AINA Workshop on Information Networking and Applications, pp 185–188, 2005.

[24] M Daoud, Q. H. Mahmoud, "Monte Carlo simulation-based algorithms for estimating the reliability of mobile agent-based systems" Journal of Network and Computer Applications, pp 19–31, 2008.

[25] ML. Shooman, "Reliability of computer systems and networks: fault tolerance, analysis, and design". New York: Wiley; 2002.

[26] C. Srivaree-ratana, A. Konak, A. E. Smith, "Estimation of all-terminal network reliability using an artificial neural network", Computers and Operations Research, Vol. 29, pp. 849–868, 2002.

IMPACT OF DIFFERENT MOBILITY SCENARIOS ON FQM FRAMEWORK FOR SUPPORTING MULTIMEDIA APPLICATIONS IN MANETs

Mohammed Saghir
Hodeidah University, Yemen

ABSTRACT

In Mobile Ad hoc Network (MANET), the mobility of nodes is a challenging issue for designers. There are lots of possibilities of mobile scenarios in this kind of network. The source, destinations and intermediate nodes may not be using the same mobile scenarios. In this study, three mobile scenarios are taken in consideration and these scenarios are source mobility, destinations mobility and intermediate nodes mobility. The impact of the three mobile scenarios on the Quality of service Multicast Framework (FQM) for supporting multimedia applications in MANETs is studied. The simulation results show that mobility of group of destinations affects the performance of FQM framework more than mobility of source. In addition, the analysis of simulation results shows that mobility of intermediate nodes does not have high effect on the performance of FQM framework when node density is not high.

KEYWORDS

MANET, Mobility, Scenarios, density& FQM.

1 INTRODUCTION

The environment for Mobile Ad hoc Networks (MANETs) is very volatile so connections can be dropped at any moment. Distant nodes communicate over multiple hops and therefore nodes must cooperate with each other to provide routing. Among types of wireless networks, MANET provides flexible communication with low cost. All communications are done over wireless media without the help of wired base stations. The challenges in MANETs are attributed to mobility of relay nodes, absence of routing infrastructure, low bandwidth and computational capacity of mobile nodes. The mobility of nodes affects the link sate and varies the number of nodes entering or leaving the neighborhood [1].

The applications of ad hoc networks are finding in several areas due to its quick and economic deployment. These applications are including military applications, emergency operations, meeting applications, law enforcement applications, collaborative and distributed applications. For military applications, mobile ad hoc networks can provide the required communication between groups of soldiers in unknown area where install fixed infrastructure may be impossible. In the emergency operations such as search and rescue, mobile ad hoc networks are very useful for establishing communication where the conventional infrastructure communications are destroyed duo to a war or earthquake. Mobile ad hoc network also useful for meeting applications where students in the class, researchers in conference or business people need to establish a meeting through voice conversation, video chatting or video conferencing. Ad hoc networks can be used to support collaborative and distributed applications where the decision of one participant depends on the current environmental conditions and on the actions of other users. An example of this type of applications is the coordination between employers in rescue agency, where the operations are based on the conditions of all affected areas. The shared characteristics of these applications are team collaboration of large number of mobile nodes, limited bandwidth, the need for supporting multimedia applications and low latency access to distributed resources as distributed database access for situation awareness in the battlefield [2]. These applications are performed in one–to–many or many–to–many communications so multicasting is very important technique for these applications.

The mobile ad hoc network is expected to be deployed in different types of environments. These environments include cities, universities, highways, markets, conferences and battlefields. The most common in these environments is the presence of obstacles that block node movement and that prevent propagation of wireless signals. Examples of obstacles include buildings, mountains, hillsides and cars.

The mobile nodes in many real life applications move in groups while others move individually and independently [3]; mobility correlation among nodes is quite common. Moreover, node mobility in real military scenarios is not always independent. In the battlefield, nodes with the same mission usually move in groups such as tank battalions or swarms in Unmanned Aerial Vehicles (UAV) networks [4].

Mobile nodes can be classified into multiple classes as walker and cars. Each class of mobile nodes has different requirements such as moving speed. In such cases, different groups of mobile nodes can be defined and only mobile nodes that belonging to the same moving speed class can merge into a group. The communications in ad hoc networks are often among teams which tend to coordinate their movements such as firemen rescue team, flood rescue team, earthquake rescue team in a disaster recovery situation and search and rescue team in law enforcement. For this, the need arises for developing efficient and realistic group mobility models. From all these requirements of classes, it is clear that mobility models are application dependent. Moreover, the various mobility models are expected to affect the performance of different MANET network protocols in different ways. Multicast protocols are being tested as they stand to get the most impact from group mobility [2].

The motivation for supporting QoS multicasting in mobile ad hoc networks is the fact that multimedia applications are becoming important for group communication. Most of the multicast applications can potentially involve in different scenarios with different mobility model dependent on the environment and the nature of the interactions among the participants in the multicast group [5].

Real-time applications over wireless ad hoc networks include video conferencing at a location without wireless infrastructure, transmitting video on the battlefield, and search and rescue operations after a disaster. Real-time applications are fundamentally different from best-effort applications, since interactive real-time applications are delay and loss packet sensitive. The later real-time packets will be dropped while best-effort packets can be accepted. Therefore, the retransmission techniques are not generally applicable to real-time interactive applications, especially in multicast situations [6]

This paper is structured as follows: Section 2 gives an overview on the previous work whereas Section 3 gives an overview on the QoS multicast framework FQM and defines the three mobile scenarios. In Section 4, the simulation results of implementing FQM with the three mobile scenarios and different node density are presented. Finally, Section 5 provides the conclusions of this study and gives some suggestions for future work.

2 Literature Review

A mobility model represents nodes distribution and movement over the network. Different studies have approved that a selection of mobile scenario can affect the performance of routing protocols in MANETs. The presence of obstacles in many mobile ad hoc network environment blocks node movement and prevents propagation of wireless signals. The people in that environment mostly travel frequently between buildings located physically close to each other while people travel less frequently to buildings further away. The traffic in that environment is concentrated in a specific area more than others. The university center is likely to be a popular destination where general services are available. Furthermore, restaurants, concerts, lectures and special events in university can all act as attraction points where students from all areas of university flow to one area at a specified time. The mobility model for that scenario is investigated and the impact on throughput and network performance resulting from such concentrated traffic areas is studied [7]. The simulation results of the obstacle model and the random waypoint model show that the two mobility models significantly impact the performance of an ad hoc network routing protocols. In addition, the results have shown that the mobility model affects a variety of characteristics, including the connectivity of the nodes and network density, as well as the packet delivery ratio and control overhead of the routing protocol.

In [8], the performance degradation due to rapidly time varying channels in a repetition based coherent cooperative system is investigated and two cooperative scenarios with static forward nodes are studied. In first scenario source node is mobile while in second scenario destination node is mobile. A detection rules is developed for a variety of mobile scenarios. The detection rules that take into account the mobility of the nodes are mostly hybrids of partially coherent detectors and non-coherent detectors. The results of implementing the two scenarios with detection rules show that source mobility affects the performance slightly more than destination mobility for both amplify and forward (AF) and demodulate and forward (DF) relays, despite the symmetry of the network.

The virtual track based group mobility model (VT) which closely approximates the mobility models in military MANET scenarios is proposed [4]. Different types of node mobility are defined when nodes are moving in group, nodes are moving individually and nodes are static. In large scale military scenarios, mobility coherence among nodes is quite common. Moreover, the VT model also models the dynamics of group mobility when mobile nodes can merge or split. The results show that performance of routing protocols under the group mobility model can be enhanced by individual nodes and static nodes. When individual nodes are randomly distributed, the connectivity among multiple groups is increased. The network performance is improved significantly in a military scenario with dominant group mobility and deploying forwarding nodes. Furthermore, the performance of mobile ad hoc network is based on the type of mobility model.

The Reference Point Group Mobility (RPGM) is introduced to represent the relationship among mobile nodes [2]. RPGM can be applied to many existing applications. Moreover, by proper choice of parameters, RPGM can be used to model several previously proposed mobility models. This study investigates the impact of the mobility model on the performance of a specific network protocols. The RPGM model applied to two different network protocol scenarios, clustering and routing. The network performance has evaluated under different mobility patterns and for different protocol implementations. The results indicate that different mobility model affect the various protocols in different ways. Furthermore, the quality of routing protocol is based on choice of mobility model.

The impact of human mobility on the link and route lifetime in mobile ad hoc network is studied and analyzed in [9]. In addition, the differences between the effect of mobility model and collisions/interference are studied. Many experiments are conducted in a typical office environment. In these experiments, twenty Personal Digital Assistants (PDAs) are distributed with IEEE 802.11b wireless interfaces to group of students and researchers to represent different test users. The users working on the same floor in a building and as result users will be in wireless transmission range most of the time. The results show that interruptions due to human mobility and collisions/interference have a completely different impact on the lifetime of links and routes. The Bypass-AODV is a new optimization of the AODV routing protocol for mobile ad hoc networks. It is proposed a local recovery mechanism to enhance the performance of the AODV routing protocol. The Bypass-AODV shows better performance than AODV with random waypoint mobility model. However, random waypoint is a simple model that may be applicable to some scenarios but it is not sufficient to capture some important mobility characteristics of other scenarios. The performance of Bypass-AODV under a different mobility models including group mobility models and vehicular mobility models is investigated in [10]. The results of simulation show a comparable performance for Bypass-AODV and AODV protocols for group mobility model while for vehicular mobility models; Bypass-AODV suffers from performance degradation in high-speed conditions.

In [11], three different kinds of node mobility situations are taken in considerations which are source node mobility, destination node mobility and whole network nodes mobility. For all these mobility situations, both reactive mode of packet transformation and proactive mode of packet transformation are taken for comparison. The three mobility situations are studied with two different Protocols namely DSR [12] and DSDV [13]. The conclusion of the work is that when source is needed to move completely, proactive mode routing (DSDV) is used. The reactive mode is preferred for the network where complete mobile network is needed. In addition, when overall delay is considered then the Proactive mode routing (DSDV) is selected.

The performance of AODV routing protocol is studied under three different mobility models [14]. In addition, a new measurement technique called probability of route connectivity is introduced. This technique is used to quantify the success rate of route established by a routing protocol. The performance of AODV routing protocol is evaluated under several link conditions. Results clearly show that mobility models affect the performance of AODV routing protocols.

The impact of mobility models on the performance of multicast routing protocols in MANET is studied and analyzed [5]. The Random Way Point, Reference Point Group and Manhattan mobility models are used as mobility models and the On-Demand Multicast Routing Protocol (ODMRP) [15][16], Multicast Ad hoc On-demand Distance Vector Routing protocol (MAODV)[17][18] and Adaptive Demand driven Multicast Routing protocol (ADMR) [19] are used as multicast routing protocols. The results of implementing three widely used mobility models and the three multicast routing protocols in NS2 have shown that different mobility models have different affect on the performance of the multicast routing protocols.

In [20], the impact of mobility predictive models on the parameters of mobile nodes such as the arrival rate and the size of mobile nodes using Pareto and Poisson distributions is investigated. The results show that when the arrival rate increases, the mobile nodes population also increases. The Pareto distribution was considered because the Poisson distribution is not accurate for the arrival distribution. In addition, the results show that the two-parameter Pareto distribution performed better than the single-parameter Pareto distribution and exponential distribution.

Previous studies have focused on the impact of mobility models on the performance of routing protocols where all network mobile nodes use the same mobility scenario. Although some previous studies [8][11] have implemented different mobility scenarios for unicast routing protocol in mobile ad hoc networks, this study focuses on the impact of different mobility scenarios for source nodes, intermediate nodes and destination nodes for multicast routing protocol in mobile ad hoc networks.

3 THE FQM WITH THE THREE CLASSES OF MOBILE SCENARIOS

In this section, an overview on the QoS Multicast Framework (FQM) is given and three classes of mobile scenarios are described.

3.1 The FQM QoS multicast framework

Multicast routing is more efficient in MANETs because it is inherently ready for multicast due to their broadcast nature that avoids duplicate transmission. Packets are only multiplexed when it is necessary to reach two or more destinations on disjoint paths. This advantage conserves bandwidth and network resources [21]. A cross-layer framework FQM is proposed to support QoS multicast applications for MANETs [22]. The FQM framework consists of five components. The first component of the framework is a new and efficient QoS multicast routing protocol (QMR) which is used to find and maintain the paths that meet the QoS requirements. The second component is a distributed admission control which used to prevent nodes from being overloaded by rejecting the request for new flows that will affect the ongoing flows. The third component is an efficient way to estimate the available bandwidth and provides the information of the available bandwidth for other QoS schemes. The fourth component is a source based admission control witch used to prevent new sources from a affecting the ongoing sources if there is not enough available bandwidth for sending to the all members in the multicast group. The fifth component is a cross-layer design with many QoS scheme: classifier, shaper, dynamic rate control and priority queue.

The traffic is classified and processed based on its priority; therefore, control packets and real-time packets will bypass the shaper and sent directly to the interface queue at MAC layer. The best-effort packets should be regulated based on the dynamic rate control. In the priority scheduling, control packet, data packets and best-effort are maintained in separate queues in FIFO order. In the scheduling algorithm, the packet with high priority should be sent firstly to the channel. Control packets should have highest priority, while real-time packets should have higher priority than best-effort packets. These schemes work together to support real-time applications.
The various mobile scenarios are expected to affect the performance of different network protocols in different ways, furthermore, the mobility models are application dependent. Three mobile scenarios are described in the following sections.

3.2 Scenario one (source mobility):

The source node in this scenario is moving according to random waypoint mobility model while destination nodes and intermediate nodes are static. The applications for sensor networks may represent this scenario. In addition, leader of groups for military in battlefield such as swarms or tank battalions also represents this scenario.

3.3 Scenario two (destinations mobility):

In this scenario, destinations are moving according to the random waypoint mobility model while source and intermediate nodes are static. Students in distance learning lectures and members in audio or video conferences may represent this scenario. The destinations may move around a static source. In addition, vehicle cars and marketing people also represent this kind of scenario.

The vehicle cars move around base stations in the road and marketing people may move around base station in the market.

3.4 Scenario three (intermediate nodes mobility):

The intermediate nodes in this scenario are moving according to random mobility models while source and destinations are static. In some video conferencing, source and destinations are static in some places and the mobility was represented by intermediate nodes that can forward data traffic without submitting to the video conferencing.

4 PERFORMANCE EVALUATION

In this section, the impact of different mobility situations on the FQM framework for supporting multimedia applications in MANETs is studied using GLOMOSIM [23]. Several mobile scenarios are used and the simulation was run using a MANET with fixed number of nodes moving over a rectangular 1000 m × 1000 m area for over 900 seconds of simulation time. In each mobile scenario, the mobile nodes are moved according to the random waypoint mobility model provided by GLOMOSIM. The Random Waypoint mobility model is flexible and can be used to create a real mobility scenario for the moving of people in many environments [24]. Mobility speed was 20 m/s and the pause time was 0 s. The radio transmission range was 250 M and the channel capacity was 2Mbit/s. Each data point in this simulation represents the average result of ten runs with different initial seeds.

The impact of different mobile scenarios on the performances of FQM for supporting multimedia applications is studied through the following performance metrics:

- *Packet delivery ratio:* the average of the ratio between the number of data packets received and the number of data packets that should have been received at each destination. This metric indicates the reliability of the proposed framework.
- *The Control overhead:* the number of transmitted control packet (request, reply, acknowledgment) per data packet delivered. Control packets are counted at each hop. The available bandwidth in MANETs is limited so it is very sensitive to the control overhead.
- *Average latency:* the average end-to-end delivery delay is computed by subtracting packet generation time at the source node from the packet arrival time at each destination. The multimedia applications are very sensitive to the packet delay; if the packet takes long time to arrive at destinations, it will be useless and will be dropped.
- *Jitter:* the variation in the latency of received packets. It is determined by calculating the standard deviation of latency. This is an important metric for multimedia applications and should be kept to a minimum value; a smaller value indicates a higher quality flow.
- *Group Reliability:* the ratio of number of packets received at 95% of destination and number of packets should be received. This means that the packet is considered to be received only if it is received by 95% of the number of multicast group.

4.1 The Performance of FQM under different mobile scenarios

In this section, the impact of the three mobile scenarios on the performance of the FQM framework for supporting multimedia applications in MANETs under different performance metrics is studied.

4.1.1 Packet Delivery Ratio (PDR)

The PDR as a function of the three mobile scenarios is given in Figure 1. The PDR for scenario 1 and scenario 3 is higher than PDR for scenario 2. This is because the mobility of group of destinations with low node density for scenario 2 affects PDR more than mobility of source in scenario 1 and mobility of intermediate nodes in scenario 3. Although, it has been proven that mobility of source affects PDR more than mobility of destination [8] and this for one source and one destination (unicast routing) whereas in these scenarios (multicast routing) group of destinations affect PDR more than one source. In addition, the PDR for scenario 3 is the highest because only intermediate nodes are mobile. The speed of mobility of intermediate nodes does not affect PDR very high because Forward Nodes (FNs) can be selected from intermediate nodes periodically every 3 seconds.

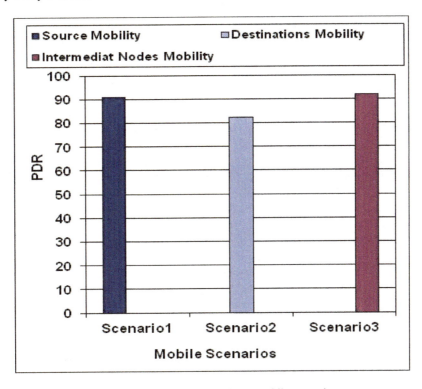

Figure 1. Performance of PDR vs. mobile scenarios

4.1.2 Control Overhead (OH)

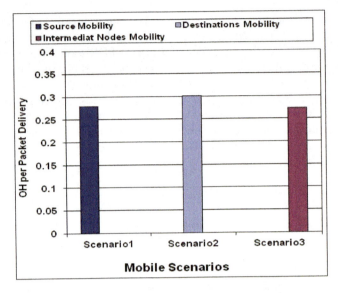

Figure 2. Performance of OH vs. mobile scenarios

The OH as a function of the three mobile scenarios is given in Figure 2 which shows that the average control overhead for scenario 2 is higher than average control overhead for scenario 1 and scenario 3. This is because the number of data packets that received at destinations for scenario 1 and scenario 3 is higher than the number of data packets that received at destinations for scenario2 as described in Section 4.1.1. In addition, the mobility in the three scenarios does not affect the control overhead because the source node sends control packets periodically every 3 seconds and as a result the number of generated control packets in the three mobility scenarios almost the same. Furthermore, the differences between the PDR in the three scenarios are reflected in the average of control OH.

4.1.3 Average Latency (AL)

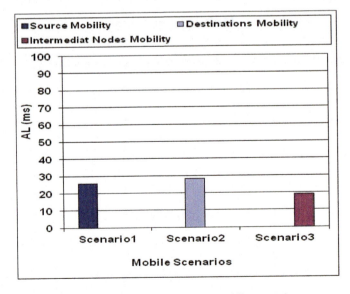

Figure 3. Performance of AL vs. mobile scenarios

The AL as a function of the three mobile scenarios is given in Figure 3. The AL for scenario 2 is higher than AL for scenario 1 and scenario 3 because the mobility of destinations in scenario 2 affects paths from source to destinations and as a result data packets take long time to arrive at destinations after destinations movements. Although source node and intermediate nodes are static, new FN nodes are needed to be selected and as a result new paths to destinations will be constructed and for this, average latency increased. In addition, the AL for scenario 3 is the lowest because source and destinations are static and as a result latency for destination nodes that received directly from source node does not affected by mobility of intermediate nodes.

4.1.4 Jitter

Figure 4 reflects the jitter as a function of the three mobile scenarios. The figure reflects that jitter for scenario 2 is higher than jitter for scenario 1 and scenario 3. This is because mobility of destinations in scenario 2 has a high effect on the constructed paths from source to destinations more than mobility of source in scenario 1 as discussed in Section 4.1.3. In addition, the jitter in scenario 3 is very low because only intermediate nodes are mobile. The mobility of intermediate nodes in scenario 3 does not have high effect on the average latency and as a result it does not have high effect on the jitter. Furthermore, the data packet latency for destination nodes that received from source directly does not change and as a result the jitter is very low.

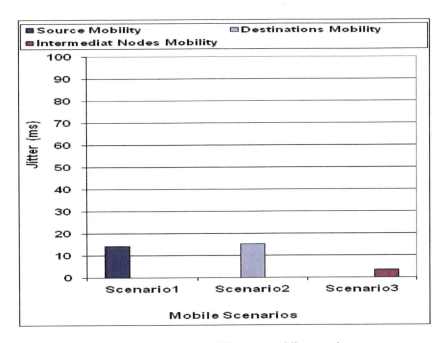

Figure 4. Performance of Jitter vs. mobile scenarios

4.1.5 Group Reliability (GR)

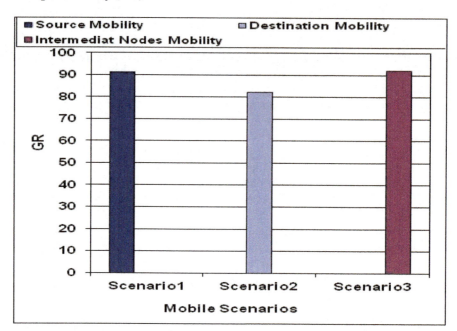

Fig.5: Performance of GR vs. mobile scenarios

The GR as a function of the three mobile scenarios is given in Figure 5. The figure shows that GR for scenario 1 and scenario 3 are higher than GR for scenario 2. As discussed in Section 4.1.1, The mobility of destinations with low node density in scenario 2 affects data packets that received at destinations and as a result some data packets are dropped before they arrive at all destinations in the multicast group. In scenario 1 and scenario 3 destinations are static and for this, data packets may have high chance to arrive at all destinations. Consequently, the GR for scenario 1 and scenario 3 are higher than the GR for scenario 2.

4.2 The Performance of FQM under different mobile scenarios and different nodes density

The mobility of nodes affects the nodes distribution based on the level of nodes density. In this section, the performance of the FQM framework for supporting multimedia applications in MANETs under different nodes density and different mobile scenarios is studied. This section focuses on the effect of the nodes density on the source mobility scenario, destinations mobility scenario and intermediate nodes mobility scenario.

4.2.1 Packet Delivery Ratio (PDR)

The PDR as a function node density for the three mobile scenarios is given in Figure 6. The PDR for scenario 3 is slightly decreased when node density increases. This is because most of nodes in the network represent intermediate nodes which are mobile. Increasing node density affects nodes distribution and increases contention and collision between mobile nodes and for this the traffic is congested and the available bandwidth is reduced. For scenario 1, when node density increased, the PDR did not affect because the mobility of source node does not affect node distribution very high. In addition, the slightly increased in PDR for scenario 2 when node density increase can be referred to the increase in the number paths from source to destinations. The differences between the PDR for three mobile scenarios with low node density (100 nodes) are discussed in section 4.1.1.

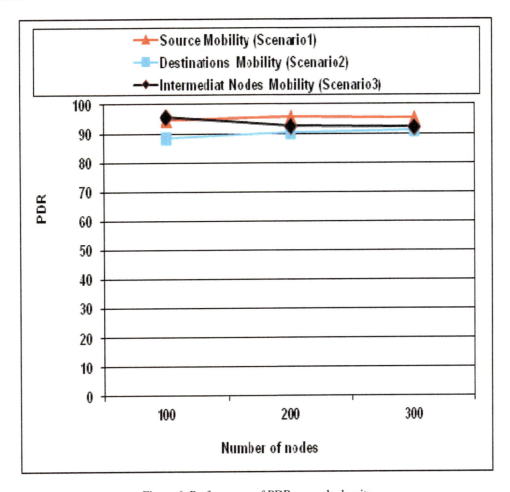

Figure 6. Performance of PDR vs. node density

4.2.2 Average Latency (AL)

The AL as a function of node density for the three mobile scenarios is given in Figure 7. The AL for scenario 2 is increased when node density increased. This is because data packets with high latency get alternative paths to arrive at destinations and as a result AL increased and PDR increased as discussed in section 4.2.1. For scenario 1 and scenario 3, the node density did not have high effect on the average latency. The differences between the AL for the three mobile scenarios with low node density (100 nods) are discussed in section 4.1.3.

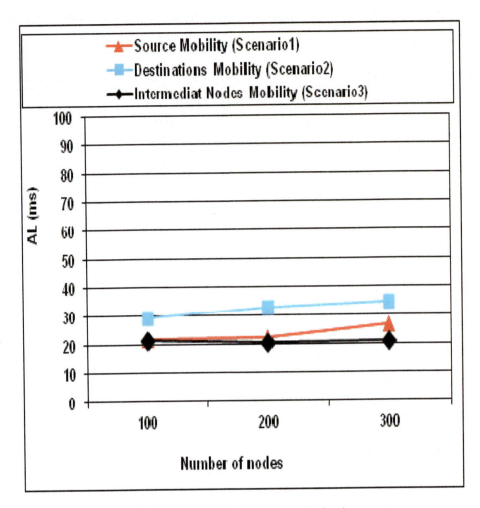

Figure 7. Performance of AL vs. node density

4.2.3 Jitter

Figure 8 reflects the jitter as a function of node density for the three mobile scenarios. The figure reflects that jitter for scenario 1, scenario 2 and scenario 3 are slightly increased when node density increased. This is because number of alternative paths increased and data packets may arrive at destinations through different alternative paths with different latency time. The differences between the jitter for the three mobile scenarios with low node density (100 nodes) are discussed in section 4.1.4.

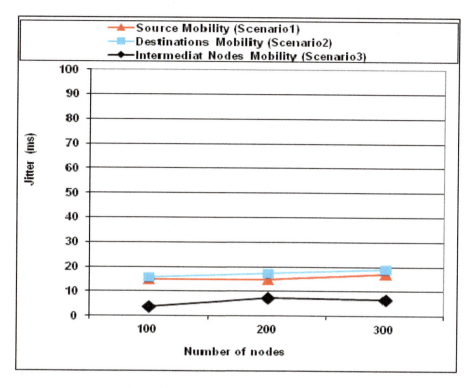

Figure 8. Performance of Jitter vs. node density

4.2.4 Group Reliability (GR)

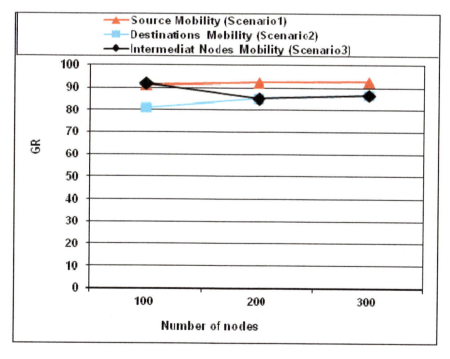

Fig.9: Performance of GR vs. node density

The GR as a function of node density for the three mobile scenarios is given in Figure 9. The figure shows that GR for scenario 3 decreased when node density increased. This is because the mobility of intermediate nodes with high node density increases contention and collision as discussed in section 4.2.1 and as a result some data packets will be dropped before they arrive at all destinations in the multicast group. The GR for scenario 2 increased when node density increased because mobility of destinations with high node density increases the number of alternative paths and for this data packets may have high chance to arrive at all destinations. The differences between the GR for the three mobile scenarios with low node density (100 nodes) are discussed in section 4.1.5.

5 CONCLUSION AND FUTURE WORK

The mobility of nodes is challenging issue for designers of mobile ad hoc network. Many types of mobile scenarios can be implemented in mobile ad hoc networks and different sets of mobile nodes in MANET may use different mobile scenarios. In this study, the performance of the FQM framework under the source mobile scenario, destination nodes mobile scenario and intermediate nodes mobile scenario is studied. The analysis of simulation results demonstrates that mobility of group of destinations scenario affects the performance of FQM framework much more than mobility of source scenario for multicast routing. Furthermore, the FQM framework can support the applications that based on source mobile scenario better than applications that based on destinations mobile scenario when node density is not high.

In addition, the mobility of intermediate nodes scenario does not have a high effect on the performance of the FQM framework either in PDR, AL, GR or jitter when node density is not high. For general, the mobility of intermediate nodes for QoS multicast protocols that use Forward Nodes does not have a high effect on the performance of the QoS multicast protocols. Furthermore, the QoS multicast protocols that use Forward Nodes to forward data packets are suitable for multimedia applications that based on intermediate nodes mobile scenario. Our future work will focus on implanting the FQM approach with different mobility scenarios under different mobility models to study its performance and efficiency.

REFERENCES

[1] N. Lakki, A. Ouacha, A. Habbani, M. AJANA EL KHADDAR, M. El Koutbi and J. El Abbadi , (2012) "A New Approach for Mobility Enhancement of OLSR Protocol", International Journal of Wireless & Mobile Networks (IJWMN) Vol. 4, No. 1, pp. 117-128.

[2] X. Hong, M. Gerla, G. Pei, and C. Chiang, (1999) "A Group Mobility Model for Ad hoc Wireless Networks", in Proceedings of the 2nd ACM International Workshop on Modeling, Analysis and Simulation of Wireless and Mobile Systems, USA.

[3] T. Ei, S. Diouba, W. Furong and I.Khider, (2008) "Autonomic Group Mobility Model for Mobile Ad hoc Networks", Proceedings of the World Congress on Engineering, London.

[4] B. Zhou, K. Xu and M. Gerla, (2004) "Group and Swarm Mobility Models for Ad Hoc Network Scenarios using Virtual Tracks", IEEE Military Communications Conference, CA.

[5] R. Manoharan and E. Ilavarasan, (2010) "Impact of Mobility On The Performance of Multicast Routing Protocols In MANET", International Journal Of Wireless and Mobile Networks (IJWMN), Vol. 2, No. 2, pp. 110-119.

[6] W. Wei and Avideh Zakhor, (2007) "Multiple Tree Video Multicast over Wireless Ad Hoc Networks", IEEE Transactions on Circuits and Systems for Video Technology, Vol. 17, No. 1, pp. 2-17.

[7] A. Jardosh, E. M. Belding-Royer, K. C. Almeroth, S. Suri, (2003) "Towards Realistic Mobility Models for Mobile Ad hoc Networks", MobiCom'03, USA.

[8] K. Srikanth and S. Ali Jafar, (2006) "Impact of Mobility on Cooperative Communication", Proceedings of Wireless Communications and Networking Conference, pp. 908-913, las vegas, NV.

[9] V. Lenders, J. Wagner and M. May, (2006) "Analyzing the Impact of Mobility in Ad Hoc Networks", Proceeding of International Work shop on Multi hop Ad hoc Networks, REALMAN'06, Italy.

[10] A. Alshanyour and U. Baroudi, (2010) "A Simulation Study: the Impact of Random and Realistic Mobility Models on the Performance of Bypass-AODV in Ad Hoc Wireless Networks", EURASIP Journal on Wireless Communications and Networking, Vol. 2010.

[11] P. Joshi, A. Gautam, A. chaudhry, P. Punia, N. Bani, (2012) "Impact of Various Mobility Model and Judgment for Selecting Mode of Network in Different Mobility Situation for Mobile Ad- Hoc Network (MANET)", 1st International Conference on Emerging Technology Trends in Electronics, Communication and Networking, Gujarat.

[12] N. Bhalaji, A. R. Sivaramkrishnan, S. Banerjee, V. sundar and A. Shanmugam, (2009) "Trust Enhanced Dynamic Source Routing Protocol for Ad hoc Networks", International Journal of World Academy of Science, Engineering and Technology Vol. 49, pp. 1074-1079.

[13] C. E. Perkins and P. Bhagwat, (1994) "Highly Dynamic Destination-Sequenced Distance-Vector routing (DSDV) for mobile computers", In SIGCOMM, pp. 234–244, ACM Press.

[14] M. Zuhairi, H. Zafar and D. Harle, (2012) "The Impact of Mobility Models on the Performance of Mobile Ad Hoc Network Routing Protocol, IETE TECHNICAL REVIEW, Vol. 29, No. 5, pp. 414-421.

[15] S.-J. Lee, W. Su and M. Gerla, (2000) "On-Demand Multicast Routing Protocol (ODMRP) for Ad Hoc Networks," Internet Draft, draft-ietf-manet-odmrp-02.txt.

[16] Mario Gerla, Guangyu Pei and Sung-Ju Lee, (1998) "On-Demand Multicast Routing Protocol (ODMRP) for Ad-Hoc Networks", draft-gerla-manet-odmrp-00.txt.

[17] E. M. Royer and C. E. Perkins, (1999) "Multicast Operation of the Ad-hoc On-Demand Distance Vector Routing Protocol," in the Proceedings of the 5th Annual ACM/IEEE International Conference on Mobile Computing and Networking, pp. 207-218.

[18] E. M. Royer and C. E. Perkins, (2000) "Multicast Ad hoc On-Demand Distance Vector (MAODV) Routing," Internet Draft: draft- ietf-manet-maodv-00.txt.

[19] J. G. Jetcheva and D. B. Johnson, (2001) "Adaptive Demand-driven Multicast Routing in Multi-hop Wireless Ad hoc Networks," In Proceedings of the ACM International Symposium on Mobile ad hoc networking and computing, pp. 33-44, USA.

[20] J. Tengviel and K. Diawuo, (2013) "Comparing the Impact of Mobile Nodes Arrival Patterns in Manets using Poisson and Pareto Models", International Journal of Wireless & Mobile Networks (IJWMN) Vol. 5, No. 5, pp. 179-187.

[21] M. Hasana and L. Hoda, (2004) "Multicast Routing in Mobile Ad Hoc Networks", Kluwer Academic Publishers.

[22] M. Saghir, T. C. Wan, and R. Budiarto, (2006) "A New Cross-Layer Framework for QoS Multicast Applications in Mobile Ad hoc Networks," International Journal of Computer Science and Network Security, (IJCSNS), Vol. 6, No.10, pp. 142-151.

[23] http://pcl.cs.ucla.edu/projects/glomosim.

[24] T. CAMP, J. BOLENG and V. DAVIES, (2002) "A Survey of Mobility Models for Ad hoc Network Research", Wireless Communications & Mobile Computing, Vol. 2,pp. 483-502.

AN EDUCATIONAL BLUETOOTH QUIZZING APPLICATION IN ANDROID

Michael Hosein and Laura Bigram

Department of Computing and Information Technology,
University of the West Indies, St Augustine, Trinidad

ABSTRACT

Bluetooth is one of the most prevalent technologies available on mobile phones. One of the key questions how to harness this technology in an educational manner in universities and schools. This paper is about a Bluetooth quizzing system which will be used to administer quizzes to students of a university. The Bluetooth quizzing application consists of a server and client mobile Android application. It will utilize a queuing system to allow many clients to connect simultaneously to the server. When clients connect, they can register or choose the option to complete a quiz that the lecturer selected. Results are automatically sent when quiz is done from the client application. Data analysis can then be done to review the progress of students.

KEYWORDS

Bluetooth, Wireless networks, Educational applications, Quizzing applications

1. INTRODUCTION

The use of Bluetooth is eminent in today's society. It is usually a default technology that is present on all mobile phones as stated by (Korucu and Alkan 2011, 1929). The ratio of mobile phones to fixed phones in Trinidad and Tobago is 6.3:1 as stated by the (International Telecommunications Union 2011). Hence, it is safe to assume that Bluetooth is one of the most prominent wireless technologies in Trinidad and Tobago.

Its uses range from transferring files to opening garage doors. (Hosny 2007, 972) stated that it is a low power, inexpensive, short-range wireless standard supporting local area networks (LANs). It is a useful tool for executing small tasks that are not data intensive.

One of the key questions is how to harness the use of this technology in the education system. The purpose of this project is to explore the use of Bluetooth as a line of communication for the delivery of educational materials. The problems that this could address at any educational institute are: 1) low attendance rate 2) low pass rates 3) poor learning curve of students.

A paper based quiz system in every class can be too time-consuming. Therefore, a Bluetooth quizzing system was thought to be the most cost effective, quick method of addressing the above problems. The system collects responses, calculate marks and send quiz scores back to students. In this way, students are encouraged to learn their work continuously before every class.

(Bar, Haussge, Robling 2007, 281) stated that when taking a 2(two) minute break after 20(twenty) minutes of lecture, the learning result of students is increased. The media break, as the study outlines, is in the form of questions directed to the students. Asking these questions, presents the students with an opportunity to reflect the learnt material. Using this concept, the report aims to utilize the Bluetooth quizzing system as the break format during lectures to stimulate the students about thinking about the course more in depth and to encourage them to ask questions. The questions in the Bluetooth quizzing system are in the form of multiple choices.

Also, (Ruhl and Suritsky 1995, 2) and (Ruhl et al. 2012, 62) indicated that the pause procedure alone was most effective for enhancing student performance on immediate free-recall of lecture ideas. Even though this study was done using students with disabilities, the same concept can be applied to any other student, with the use of quizzes during lectures.

The system consists of a server component and a client component. The server component is controlled by the lecturer whereas the client component is in the hands of the students. Both components are in the form of an Android application. At the server, the lecturer allows students to register (send their initial data to the server), add/edit courses, students and quizzes. Registration consists of data such as first name, last name student identification number among other bio-data. The lecturer can also create quizzes consisting of multiple choice questions. After creating these quizzes, the lecturer can dispense the quiz to students via the quizzing phase (This is explained in greater detail later). The quizzing phase also consists of sending back results to students. In addition, the lecturer can then perform data analysis on the data received. Statistics is represented with the use of pie charts. This will allow the lecturer to keep track of the students' performances. This Bluetooth quizzing system will be referred to as BLUEQ.

2. LITERATURE REVIEW

There have been other attempts to implement a classroom Bluetooth quizzing system to encourage interaction between the lecturer and students.

(Davidrajulh 2009) presented a Bluetooth-based classroom tool. It is a paper focused on evaluating a Bluetooth-based classroom tool. This tool was used to help lecturers automate their assignment tests. The paper however, was limited to 2 (two) handhelds and a master device. In this system, students used their mobile phones to submit answers to assignment tests to the lecturer's computer via Bluetooth. By doing this, the lecturer does not need to correct assignment tests. In addition, the system was also used to disperse multiple choice questions to the students during the lectures. The students would then submit their answers to the lecturer's machine. In both scenarios however, no test scores were sent back to the student.

Another system was also developed by (El Sharkawy and Meawad 2009) which was a mobile quizzing system through which students can answer short questions in lectures with the use of SMS (short messaging service) or Bluetooth. The main objective of this system was to provide the students with different technology options that would enhance their learning experience, as well as, encourage them to attend lectures thereby increasing the level of participation. The server component of this system consisted of a web module, a GSM modem, and SMS gateway and a Bluetooth module.

The client consisted of Java ME mobile application. The evaluation of this system showed that students were enthusiastic about using the system. The system also consisted of a statistical section that allowed the lecturer to view the statistics after a quizzing session. In this section, all the submitted answers are gathered, and charts displaying the different answers were shown.
Another complex system was developed by (Bar et al. 2006, 361) which encompassed SMS (short messaging service), WLAN (wireless local area network) and Bluetooth as part of the system to

engage the students in their classrooms (Figure 1 shows the arrangement).It allowed students to use their mobile devices, using Bluetooth to interact with the educator during lectures. That is, it allowed students to answer questions presented by the educator. The answers to the questions were presented at the end of the lecture. Their Bluetooth model was designed for Linux using Blueproxy (Blueproxy is a simple proxy server to convert Bluetooth RFCOMM connections into TCP connections). While this was a fantastic idea for implementing the system, this option is costly (SMS is costly) whereas Bluetooth is free and pervasive. Hence, Bluetooth will be a better suited option for the delivery of content to students via the Bluetooth quizzing system. However, SMS is facilitated once in close proximity to a cell tower, whereas Bluetooth has a range of 10 meters for class 3 devices.

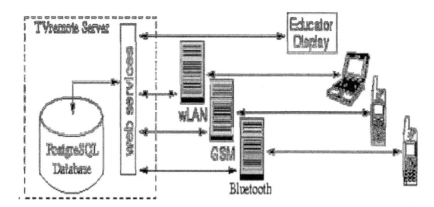

Figure 1 – The system architecture (Bar et al. 2006).

Yet another (Zhang, Li, and Fu 2007) Bluetooth based e-learning system, was used to break the constraints of time and space (The teacher received instant feedback from students and check students' performance at any time. Later (Zhang, Xiong, and Luo 2011) applied the same principle to a mobile English assistant learning system based on Bluetooth. That is, the system was later built around an English lecture session.

(Mitchell et al. 2006) investigated the use of mobile and smart phones as a platform for delivering mobile learning services and administrative information on a personalized basis. The system utilized two technologies SMS (short messaging service) and Bluetooth. The two technologies were used to complement each other to offer an alternative communications platform for students. This combination provided a mechanism for communication with undergraduates on a large scale. Figure 2 shows the architecture, with the flow beginning the mobile phone.

Figure 2 – The system architecture for (Mitchell et al. 2006)

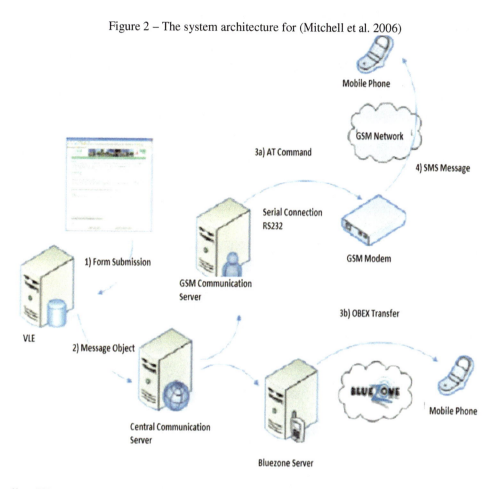

Finally, (Wang, Zhu, and Zhao 2010) developed a wireless communication educational lab based on Bluetooth. The goal of this system was to encourage students to learn by themselves. This lab focused on practical assignments which would help the students develop skills not found in a textbook. The aim was to help students grasp wireless communication from theory to practice.

As it can be seen, there have been many attempts at creating a Bluetooth quizzing application for interaction between lecturer and student. However, these implementations are a bit outdated and none of have been designed for the Android platform, which is the leading operating system today. According to International Data Corporation (2013), the Android operating system (OS) had 68.8 % market share, which makes it the best OS to create applications that will reach a wider audience (in this case students). This report seeks to fill this gap while solving the issues outlined in the Introduction.

3. APPLICATION DETAILS

There are 2 components to the Bluetooth quizzing system: Server Side and Client Side. Both components are implemented on the Android operating system (on a phone). Android was chosen as it is very popular mobile platform today. Options presented at the server are: Start Server, Manager, Data Analysis and Exit. Figure 3 shows the main menu for the server. 'Exit' is used to exit the application. 'Start Server' turns the server on/off, monitors incoming connections as well as provides a list of quizzes to allow the lecturer to select a quiz to send to students. 'Manager' consists of creating and editing new courses, quizzes and students. 'Data Analysis' is for performing basic statistical analysis on results obtained from quizzes.

Figure 3 – The main menu for the server application of BLUEQ.

Figure 4 – Client/Server sequence diagram.

Server/Client Bluetooth Communication

```
Server                                    Client
  |                                         |
  |  startServer()                          |
  |←—┐                                      |
  |  |                                      |
  |                   bluetoothOn()         |
  |                                    ←—┐  |
  |                                       | |
  |  ensureDiscoverable()                   |
  |←—┐                                      |
  |  |                                      |
  |—————————————┐                           |
  | Calls BluetoothQuizService start()      |
  |                                         |
  |  setpService()                          |
  |←—┐                                      |
  |                                         |
  | Listen for clients in thread            |
  |  run()                                  |
  |←—┐                                      |
  |                                         |
  |                   scanForDevices()      |
  |                                    ←—┐  |
  |                                         |
  |                   Starts discovery devices
  |                                         |
  |                   List of devices will be shown.
  |                   The user will then choose the Server.
  |                                         |
  |                   findingConnection()   |
  |                                    ←—┐  |
  |                                         |
  |                   Finding rfcomm channel to server
  |                                         |
  |             connect()                   |
  |  ←——————————————————                    |
  |             accept()                    |
  |  ——————————————————→                    |
  |                                         |
  | Spawn BluetoothSocket                   |
  | when client connected                   |
  |                                         |
  |             connected()                 |
  |  ——————————————————→                    |
  |                                         |
  |             startComms()                |
  |                                    ←—┐  |
  |                                         |
  |                   On Accepting input and output streams and
  |                   then established. RFCOMM
  |                   communication begins.
  |                                         |
  | then listens                            |
  | on another RFCOMM channel for           |
  | another incoming client                 |
  |                                         |
Server                                    Client
```

In Figure 4, the server is started when the user presses the Start Server toggle button on the server application. Figure 5 shows the "Start Server" screen of the BLUEQ application.

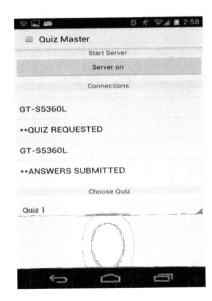

Figure 5 – Start Server screen of the BLUEQ application.

The method start Server() turns on Bluetooth at the same time. In order for client devices to find the server, it must be made discoverable by calling ensure Discoverable(). Once the server can be seen, then the quizzing service is started by calling setup Service(). At this point, the server is started and sequentially selects a UUID from a queue of UUID's. A Bluetooth Server Socketis created listening on the RFCOMM channel associated with this UUID. When a client opens a connection using the same UUID, a Bluetooth Socket is spawned for that client. The server then listens on another RFCOMM channel for another client.

A blocking queue was used to store the UUIDS that are going to be used to get the RFCOMM channels. There are 8 UUIDS that will be selected. Since the accept() method is a system blocking call, it states that the device will not be able to perform anything else, hence only one server can be used. But the UUID's will be cycled to choose the correct RFCOMM.

According to the RFCOMM white paper, protocol theoretically supports up to 60 simultaneous connections between two Bluetooth devices. The number of connections that can be used simultaneously in a Bluetooth device is implementation-specific. In experiment at with the android devices, this value was found to be roughly 5. More memory could contribute to more connections simultaneously. Hence, we estimated around 8 UUID's, that's why 8 UUIDswere chosen.

For the client application to work, the Bluetooth must be turned on. The client has to then search for the server by doing a device discovery. Scan For Devices () searches for all devices in close proximity to the client. A list of devices found is presented to the user. User interaction is required to select the server.

Once the server is selected, a service discovery is executed. This is where the client uses the same UUID as the server. The server then does a look in the service discovery database to ensure service is listed. Once the match is successful, the server sends the RFCOMM channel number on which the service is listening.

The client then connects to the server using the RFCOMM channel number. Data transfer to the server then begins. Data transfer in this case, will be the requests sent by the client. For example, as in Figure 6 showing the client requested a quiz, which is shown.

Figure 6– Quiz details presented to the student.

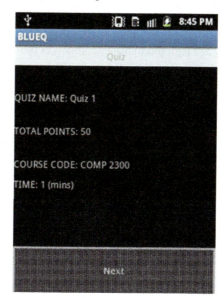

4. MODELING

This section presents a probability model of how successful clients will be in connecting to the server. We have 8 UUIDS, assuming there n = 25 (25 clients trying to connect).

We have managed to model the above scheme using the geometric distribution.

X is the discrete random variable, which is the number of attempts needed to connect to an RFCOMM channel successfully.

X~Geo(p) where p = 8/25 = 0.32(assuming 8 UUIDs always in the queue in the client, since only 1 can connect at a time)and q = 17/25=0.68

Probability that the first success is obtained at the rth attempt $P(X=r) = q^{r-1} \times p$

So finding the probability that the client connects on its 2th attempt, $P(X=2) = (17/25)^1 (8/25) = 0.22$

And finding the probability that the client connects on its 6th attempt, $P(X=6) = (17/25)^5 (8/25) = 0.05$

Hence, the probability that the client takes a large number of attempts to connect decreases as the number of attempts increases. This means the probability of connecting on the first few attempts is large.

5. CONCLUSION

A Bluetooth quizzing application was developed for use in lectures, to help lecturers administer quizzes and in turn help students revise work previously learnt. Students will benefit by revising work done in previous classes.

The queuing method of UUID's proved to be very useful due to the constraints of the Bluetooth technology. The probability of taken too many attempts was found to be very small.

Future works could improve on this method for supporting more simultaneously Bluetooth connections on the server side. Also, the means of communicated can be improved by using XML to structure messages passed between client and server.

REFERENCES

[1] Android Developer. 2013. "Bluetooth Adapter". Accessed June 11, 2013. http://developer.android.com/reference/android/bluetooth/BluetoothAdapter.html.
[2] Atmel Coporation .1991.The Bluetooth Wireless Technology. Atmel Corporation
[3] Bär, Henning, Gina Häussge, and Guido Rößling. 2007. "An Integrated System for Interaction Support in Lectures." ACM SIGCSE Bulletin 39 (3): 281. doi:10.1145/1269900.1268865.
Bär, Henning, Guido Rößling, Erik Tews, and Elmar Lecher. 2006.
[4] "Bluetooth Interaction Support in Lectures." In Proceedings of Mobile Learning, 2006, 360-364.IADIS Press.
[5] Bluetooth Special Interest Group. 2009. Specificiation of the Bluetooth System, Core Version 3.0. Accessed August 24, 2012. https://www.bluetooth.org/en-us/specification/adopted-specifications
[6] Davidrajuh, Reggie. 2009. "Evaluating Performance of a Bluetooth-Based Classroom Tool." International Journal of Mobile Learning and Organisation 3(2): 148-163. doi: 10.1504/IJMLO.2009.024424.
[7] Zhang, Guoliang, Ningbo Univ, Ningbo, FengXiong, and Qi Luo. 2007. "Mobile English Assistant Learning System Based." In 2nd International Conference on Pervasive Computing and Applications, 26-27 July, 2007, 689 - 692. Birmingham.
[8] Hosny, W. 2007. "Power Engineering Mobile Education Technology." 2007 42nd International Universities Power Engineering Conference, 4-6 September, 2007, 971–974. Brighton.
[9] Hopkins, Bruce, and Ranjith Antony. 2003. Bluetooth for java .Apress.
[10] Huang, Albert S., and Larry Rudolph. 2007. Bluetooth essentials for programmers.Cambridge University Press.
[11] International Telecommunications Union. 2011. "Mobile Cellular telephon subscriptions." Accessed May 24, 2013.http://www.itu.int/ITU-D/icteye/Reporting/ShowReportFrame.aspx?ReportName=/WTI/CellularSubscribersPublic&ReportFormat=HTML4.0&RP_intYear=2011&RP_intLanguageID=1&RP_bitLiveData=False.
[12] International Data Corporation. 2013. "IDC – Press Release." Accessed May 24, 2013. http://www.idc.com/getdoc.jsp?containerId=prUS23946013#.UTCOPjd4DIY
[13] Miller, Michael. 2001.DiscoveringBluetooth.Sybex Incorporated.
[14] Mitchell, Keith, Nicholas P. Race, Duncan McCaffery, Mark Bryson, and Zhen Cai. 2006. "Unified and Personalized Messaging to Support E-Learning." 2006 Fourth IEEE International Workshop on Wireless, Mobile and Ubiquitous Technology in Education,16-17 November, 2006, 164–168. Athens.
[15] Kammer, David, Gordon McNutt, Brian Senese, and Jennifer Bray. 2002. Bluetooth application development guide: The short range interconnect solution. Syngress Publishing Incorporated.
[16] Korucu, AgahTugrul, and AyseAlkan. 2011. "Differences Between M-learning (mobile Learning) and E-learning, Basic Terminology and Usage of M-learning in Education." 3rd World Conference on Educational Sciences, 2011,1925-1930.
[17] Kumar, C. Bala, Paula J. Kline, and Timothy J Thompson. 2004.Bluetooth application programming with Java API. Elsevier Incorporated.

[18] Rößling, Guido, Ari Korhonen, Rainer Oechsle, J. Ángel Velázquez Iturbide, Mike Joy, Andrés Moreno, AtanasRadenski, et al. 2008. "Enhancing Learning Management Systems to Better Support Computer Science Education." ACM SIGCSE Bulletin 40 (4): 142. doi:10.1145/1473195.1473239.

[19] Ruhl, Kathy L, Charles A. Hughes, Anna H. Gajar, Kathy L Ruhl, Charles A. Hughes, and Anna H. Gajar. 2012. "Efficacy of the Pause Procedure for Enhancing Learning Disabled and Nondisabled College Students ' Recall Long- and Short-Term Facts Presented Through Lecture." Learning Disability Quarterly 13 (1): 55–64.

[20] Ruhl, Kathy L, and Sharon Suritsky. 1995. "The Pause Procedure And / Or An Outline : Free Recall Effect On Immediate And Lecture Notes Taken by College Students with Learning Disabilities." Learning Disability Quarterly 18 (1): 2–11.

[21] Sharkawy, Bahia Fayez El, and FatmaMeawad. 2009. "Instant Feedback Using Mobile Messaging Technologies." 2009 Third International Conference on Next Generation Mobile Applications, Services and Technologies, September: 539–544. doi:10.1109/NGMAST.2009.93.

[22] Wang, Qing, Xiuxin Zhu, and Gaoxing Zhao. 2010. "Wireless Communication Educational Lab Construction Based on Bluetooth." 2010 5th International Conference on Computer Science & Education. August: 1574–1577. doi:10.1109/ICCSE.2010.5593770.

[23] Zhang, Yanhui, Wu Li, and Yingzi Fu. 2007. "A Mobile Learning System Based on Bluetooth." Third International Conference on Natural Computation, August: 768–771. doi:10.1109/ICNC.2007.64.

[24] Zhang, Yonghong, Shiying Zhang, Son Vuong, and Kamran Malik. 2006. "Mobile Learning with Bluetooth-based E-learning System." Proceeding of the 2005 2nd International Conference on Communications and Mobile Computing – IWCMC, 15-17 November, 2005, 5. Guangzhou.

A NOVEL APPROACH FOR MOBILITY MANAGEMENT IN LTE FEMTOCELLS

[1]Pantha Ghosal, [2]Shouman Barua, [3]Ramprasad Subramanian, [4]Shiqi Xing and [5]Kumbesan Sandrasegaran

[1,2,3,4,5,6]Centre for Real-time Information Networks

School of Computing and Communications, Faculty of Engineering and Information Technology, University of Technology Sydney, Sydney, Australia

ABSTRACT

LTE is an emerging wireless data communication technology to provide broadband ubiquitous Internet access. Femtocells are included in 3GPP since Release 8 to enhance the indoor network coverage and capacity. The main challenge of mobility management in hierarchical LTE structure is to guarantee efficient handover to or from/to/between Femtocells. This paper focuses, on different types of Handover and comparison performance between different decision algorithms. Furthermore, a speed based Handover algorithm for macro-femto scenario is proposed with simulation results

KEYWORDS

Femtocell Access Point (FAP), Handover Hysteresis Margin (HMM), Reference Signal Received Power (RSRP), Reference Signal Received Quality (RSRQ), Signal to Interference Plus Noise Ratio (SINR, Evolved NodeB (eNB), User equipment (UE).

1. INTRODUCTION

In the next generation wireless communication systems, the primary challenge is to improve the indoor coverage, capacity enhancement as well as to provide users the mobile services with high data rates in a cost effective way. Performance of mobile system can be enhanced by evolving to emerging broadband technologies such as WiMAX [1] and LTE [2] but this may not be able to endure the exponential rise in traffic volume. These advancements in 4G physical layer (PHY) are approaching to the Shannon limit [3] and ensure maximum achievable data rate. So, further enhancement either in the PHY layer or available spectrum will not be adequate to overcome the coverage and capacity challenge. One of the approaches of solving this capacity and coverage related problem includes moving the transmitters and receivers closer to each other. This method loses its ground on economic feasibility because of deploying more base stations (BSs). Thus, small cells generally known as Femtocells with restricted access permission to fewer users compared to macrocell are chosen by the mobile operators as a possible solution to improve the network coverage, especially to the indoor users with ubiquitous high speed connectivity. These Femtocell base stations are referred to as Femto Access Points (FAPs). They have a short-range (10-30m) and require a low power (10-100mW) [5] to provide high-bandwidth wireless communication services in a cost effective way. Femtocells incorporated with the plug and play capabilities work in mobile operator owned licensed spectrum and enable Fixed Mobile Convergence (FMC) [6] by connecting to the core network via broadband communications links (e.g., DSL). Unlike macrocells, FAPs are typically installed and maintained by the end users in an unplanned manner and don't have X2 interface between them for information sharing. Due to this uncoordinated nature femtocell pose challenge on Handover and Radio Resource Management. The rest of the article is organized as follows: Section.2 describes the LTE Network architecture with Femtocells, Section.3 depicts the HO types and open challenges in HO management, Section 4. Describes different HO algorithms and their performance comparison and in Section. 5 proposed HO algorithm with simulation result is described.

2. LTE NETWORK ARCHITECTURE

The 3GPP, LTE is a packet-switched with flat architecture and is mainly composed of three parts: the UE, the e-UTRAN, and the packet switched core network (EPC). EPC is responsible for all services provided to UE including voice and data to the user using packet switching technology. e-UTRAN has only one node i.e., the evolved NodeB (eNB) which handles the radio communication between UE and EPC. The physical layer of radio interface supports both time (TDD) and frequency (FDD) division duplexing. On the other hand, Femtocells not only boost indoor coverage [9] and capacity but also improve battery life of UEs since UE doesn't need to communicate with a distant macrocell base stations. Fig. 1(a) shows the basic two tier macro-femto network architecture and Fig. 1(b) shows X2 and S1 interfaces. FAPs which have a less computational capability [3], are connected through DSL (Digital Subscriber Line) in indoor scenarios. The LTE macro system based on flat architecture connects all the eNBs through X2 interface and the RRM procedure is done by eNB.

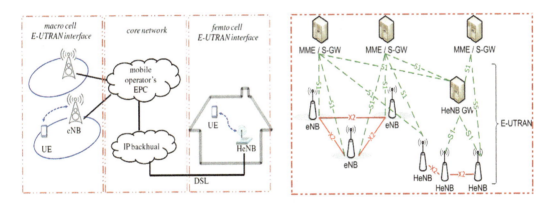

Figure 1:a) Two-tier macro-femto network architecture [3], b) Network Interfaces[9]

Femtocells can to operate in one of three access modes, i.e., closed access mode, open access mode or hybrid access mode [7]. Closed access mode is generally deployed in residential scenarios and a group of registered users called Closed Subscriber Group (CSG) have the permission to access the femtocell. In case of open access mode, any UE can access the femtocell and benefit from its services but when it comes to resource usage, congestion and security, open access is not a suitable choice. In hybrid access mode, a limited number of unknown MUE may access the femtocell while a fixed number of users defined by the owner can access the femtocell ubiquitously but may suffer the risk of security breach [8]. In this article, Closed Access Mode is considered because of security and privacy of the owner.

3. Handover to Femtocells and Challenges

In the two tier macro-femto scenario there are three possible handover scenarios as shown in Fig.2. When a UE is moving in femtocell coverage from macrocell coverage the HO that takes place is called Inbound HO and Outbound HO is one where UE gets out from the femtocell coverage to the macrocell coverage. Femtocell user (FUE) moving one FAP coverage area to another FAP coverage area is called Inter-FAP HO. LTE doesn't support X2 interface between eNB and FAP and the large asymmetry in received signal strengths makes inbound and inter-FAP HO more complex. While inbound HO, apart from received signal strength and signal quality access control, interference, user speed and position has to be taken in consideration. On the other hand, since UE moves from femtocell coverage to the macrocell coverage stored in its neighbour list with best received signal strength, the outbound HO procedure is not that complicated. The HO phase for two-tier system can be divided in to six phases[9] : 1)Cell identification 2)access control 3) cell search 4) cell selection/reselection 5)HO decision 6) HO execution. The HO phases are shown in Fig.3. The position of the femtocells are known since they are connected to the network through backhaul and the whitelist of accessible FAPs are stored in user U-SIM[9]. When user comes near to FAP coverage it gets a proximity notification from the network and collect signal information according to eNB prescribed measurement configuration. FAPs are identified through their physical cell id(PCI). The number of PCI is limited, which is only 504. So, in case of unplanned and dense deployment of FAPs, cell selection/resection/search it may become confusing to choose the accessible FAP when there are more than one FAP with the same PCI. Unable to perform that may increase the number of HO failure [9]. As it can be seen from Fig. 3 with the presence of Home eNB gateway (which is used for UE authenticate and access control) two additional steps are taken while HO decision and Execution. The additional steps as illustrated in Fig. 3($\oplus$ sign) increase the delay in connection setup.

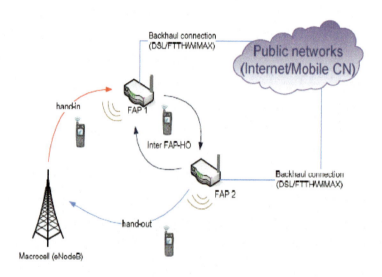

Fig. 2: Handover Scenario in presence of FAP [14]

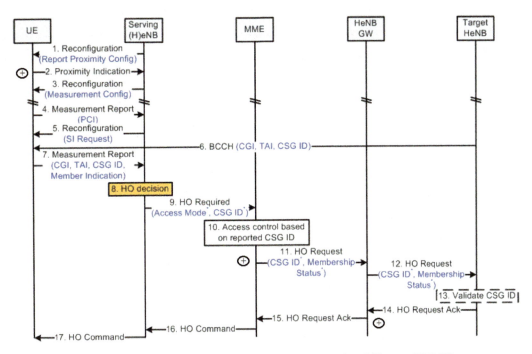

Fig. 3: HO execution signalling procedure for inbound mobility to a FAP [9]

4. Handover Decision Algorithms

The Handover decision criteria form macro-femto two tier network is different than the macro cellular network since there is no direct interface like X2 between them. In this paper, different proposed HO algorithms based on different parameters are discussed. The main decision parameter for handover to/from/between FAPs [9] can be divided in in five groups: Received signal strength (RSS/RSRP), b) User speed, c) cost-function based, d) Interference experienced at user end or serving cell, and e) Energy Emission. Since, FAPs have less computational capability and also prone to delay and congestion due to external backhaul (DSL), in this paper we have considered received signal strength, speed and interference for evaluation and evolving proposed HO algorithm.

4.1.1 Received Signal Strength Based HO Algorithm

The handover decision algorithms in this class are based on the Received Signal Strength as shown in Fig. 4. To minimize the HO probability and ping-pong effect the RSS based algortihms consider a HO Hysteresis Margin (HMM) to compare the received signal strength of the source and target cells.

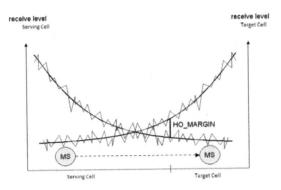

Fig. 4: Conventional Received Signal Strength Based Handover

When an UE enters the coverage area of femtocells it experiences uneven transmission powers from macrocell and femtocell. The proposed algorithm in [10], Compensates the uneven RS power transmission in single macrocell-femtocell scenario by combining RSS of the macrocell and femtocell. This algorithm uses an exponential window function to mitigate the first variation of RSS. The operation can be expressed as follows:

$$\bar{s}_m[k] = w[k] * s_m[k] \quad (1)$$
$$\bar{s}_f[k] = w[k] * s_f[k] \quad (2)$$

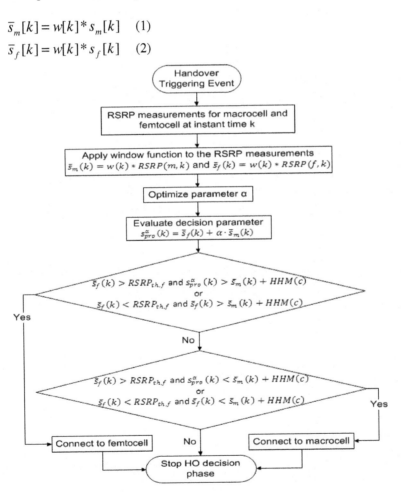

Fig. 5: HO Algorithm based on RSS [10]

Here, $w(k)$ denotes the exponential window function and $\overline{s}_m[k]$, $\overline{s}_f[k]$ represents the filtered RSS parameters of the macrocell and the femtocell at time k respectively. These filtered signals are then combined into a RSS-based decision parameter as follows:

$$s_{pro}^{\alpha}[k] = \overline{s}[k] + \alpha \overline{s}_m[k] \qquad (3)$$

Where $a \in [0, 1]$ is the combination factor to compensate the large asymmetry between transmit power of eNB ($\approx$46dBm) and HeNB($\approx$20dBm). The algorithm proposed in [10] is depicted in Fig.5. For inbound mobility HO to the femtocell is possible,

$$\text{If, } \overline{s}_f[k] > s_{f,th} \text{ and } s_{pro}^{\alpha}[k] > \overline{s}_m[k] + HMM \qquad (4)$$

$$\text{or} \quad \text{if, } \overline{s}_f[k] < s_{f,th} \text{ and } \overline{s}_f[k] > \overline{s}_m[k] + HMM \qquad (5)$$

On the other hand, for connecting to macrocell from femtocell is made,

$$\overline{s}_f[k] < s_{f,th} \text{ and } \overline{s}_f[k] + HMM < \overline{s}_m[k] \qquad (6)$$

$$\text{or} \quad \text{if, } \overline{s}_f[k] > s_{f,th} \text{ and } s_{pro}^{\alpha}[k] < \overline{s}_m[k] + HMM \qquad (7)$$

The advantages of this algorithm are, it considers the asymmetry in transmit powers between eNB and HeNB and it also includes optimization parameter for the trade-off between HO probability and number of HO failure. Nevertheless, UE speed, bandwidth availability, user subscription and interference were not considered in this single macro-femto model

4.1.2 Received Signal Strength and Path Loss Based HO Algorithm:

The proposed HO decision algorithm in [11] considers path loss along with RSS for inbound mobility to femtocells. Similar to the proposals in [10], this path-loss based algorithm also considers exponential window function $w(k)$ on the RSRP measurements to compensate the asymmetry level between macro-femto RS transmission powers. Handover to femtocell from macrocell is possible if a) the filtered RSRP measurement of the femtocell exceeds over a minimum threshold, denoted by $RSRP_{th,f}$, b) the filtered RSRP status of the femtocell exceeds over the filtered RSRP status of the macrocell plus the HHM, and c) the observed path-loss between user and FAP is less than the path-loss between UE and the macrocell. The HO algorithm flowchart is illustrated in Fig.6 [11].

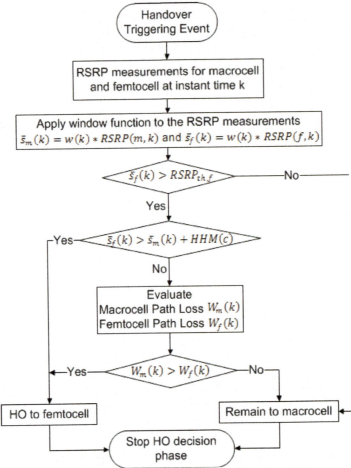

Fig. 6: HO Algorithm based on RSS and Path loss [11]

The main feature of this algorithm is that, it considers actual path-loss between UE and the target cell. However, the algorithm considers single macro-femto model which may not be realistic. On the other hand, path-loss is prone to fast variation which will in turn influence ping-pong effect while HO.

4.2 Speed based HO Algorithm

The main decision criterion for this type of algorithm is speed. In [12], authors proposed HO algorithm conceiving two decision parameters, speed and traffic type. Based on speed either proactive or reactive HO decisions are performed. Proactive HO is one where HO takes place before RSS of the serving cell reaches a pre network defined hysteresis margin. In this type of HO strategy a residual time prior HO execution is estimated. To minimize the HO delay and packet loss for real time traffic is the purpose of pro-active handover. In reactive handover, HO is execution is initiated when minimum required RSS is reached. The purpose of reactive HO is to reduce ping-pong effect. Fig.7 illustrates the operation of

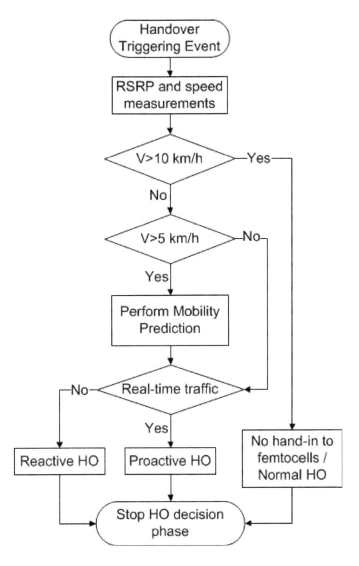

Fig. 7: Speed based HO algorithm [12]

multiple macrocell-femtocell scenarios. In the below figure, when the UE speed is higher than 10kmph, there will be no HO from macro to femtocell. When UE has the speed between 5 to 10kmph then this algorithm performs prediction model using Markov-Chain [13] to predict the direction of the user using current position and speed. If the UE moves towards the femtocell, then the proposed model performs either proactive HO if the traffic is real time or reactive HO if the traffic is non-real time. Same approach is made for the users below 5kmph without mobility prediction. This proposed algorithm expected to reduce the HO probability for the users with medium speed (5kmph<speed<10kmph) and better QoS for the real-time traffic users.

5. A Novel Speed Based Algorithm

Femtocells are connected to network through backhaul broadband connection. Due to the absence of X2 interface, the HO decision and execution phase take more time than the conventional macro-cellular handover. Moreover, FAPs have less computational capability. But HO decision algorithms based on energy efficiency [16, 17] or cost function [15] include complex algorithms for FAP to manage and it will also delay the HO decision procedure. Keep these factors in mind;

we tried to build up a simple speed based HO algorithm that also includes the other handover parameters i.e., a) access control b) bandwidth satisfaction c) received signal strength and d) traffic type. Our model shown in Fig. 8 is inspired by speed based HO algorithm proposed in [12] .Unlike the model [12], we didn't consider Markov-Chain prediction model since MME knows the speed and location of user [18], which makes the HO decision simple and faster.

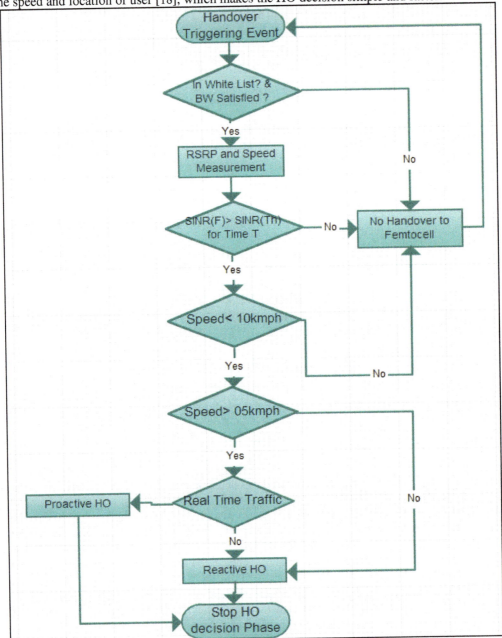

Fig. 8: Proposed Speed Based Handover Model

We used LTE-Sim (an event driven object oriented simulator written in C++ [19]) for simulating inbound Handover. Different packet scheduler (PF, EXP-PF, MLWDF, FLS) are available in LTE-Sim module eNB to perform data flow and resource allocation. The performance of the packet schedulers (PF, M-LWDF and EXP/PF) considering all the users are experiencing single flow (50% of the users are having VoIP flow and the rest are having Video flow) modelled with infinite buffer application was measured prior simulating the proposed algorithm. Fig. 9(a), 9(b) and 9(c) shows the performance of three different packet schedulers in terms of Throughput, Packet Loss Ratio (PLR) and Spectral efficiency considering users are moving with the speed of 3Kmph (Pedestrian Speed) and 120Kmph (Highway Speed). While simulating Random walk model [20] was considered. All the results show in all cases in LTE-Sim, MLWDF has the better performance. The simulation parameters are mentioned in Table-1.

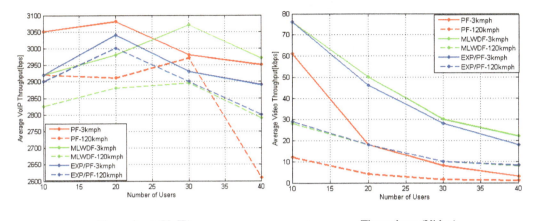

Throughput (VoIP)   Throughput (Video)

Fig. 9 (a) Average Throughput of **VoIP** and **Video Flow** with different schedulers at different speed [21]

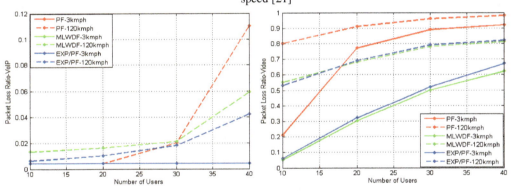

PLR (VoIP)   PLR (Video)

Fig. 9 (b) Packet Loss Ratio (PLR) of **VoIP** and **Video Flow** with different schedulers at different speed [21]

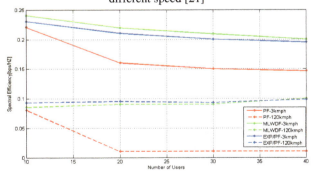

Fig. 9(c) Spectral Efficiency of different schedulers in LTE-SIM [21]

For the simplication of the HO algorithm, PF was considered in our proposed HO algorithm. Fig. 10 shows the comparison between the assignment probabilities of our proposed model to the earlier mentioned models of RSS based HO model [11] and speed based HO model [12]. From Fig. 10 it can be seen that, our proposed HO model is showing better performance when FAP is located near the eNB and the assignment probability to FAP is higher than other two models.

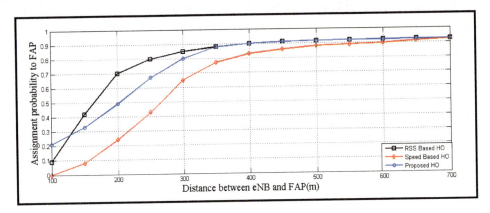

Fig. 10: Assignment Probability to FAP vs. distance between eNB and FAP

Table 1: Simulation Parameters.

Total Bandwidth	20 MHz
eNb power transmission	43 dBm
FAP power transmission	20 dBm
CQI	Full Bandwidth and periodic (2ms) reporting scheme
Apartment Size	100 m^2
Building Type	5*5 Apartment grid
Number of FAPs	1 FAP/Apartment
CSG Users	9 FUE/FAP
Scheduler	PF
Traffic	VOIP, VIDEO
Mobility	Random Walk Model

In Fig.11 it can also be seen that, in our proposed model the number of handover is less than RSS based or Speed based HO models within 350 meters. But after 350 meters it shows higher handoffs than the RSS based HO model, because at the cell edge the downlink interference experienced because of the presence of MUE near FAP is higher.

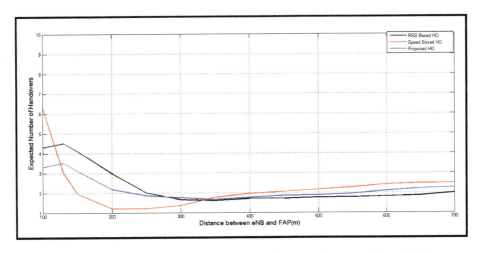

Fig. 11: Number of Expected Handovers vs. Distance between eNB and FAP

Table 2: Features of the Proposed Speed Based HO Algorithm

Signal Strength	Strengths	Future Modification
Minimum RSS for HO	√	
Path Loss	√	
Window Function	√	
SINR	√	
Speed		
UE Speed	√	
UE mobility Pattern		X
BW Related		
Cell Capacity	√	
Cell Load	√	
Number of UE's Camped		X
Cell Type	√	
Traffic Related		
Traffic Type	√	
BER		X
Energy Efficiency Related		
UE power Class		X
UE battery Class		X
Mean UE transmit Power		X
Access		
UE membership Status	√	
UE Application priorities	√	
UE Priorities		X

6. CONCLUSION

In our paper, we tried to show the mathematical complexity of the renowned approaches for HO techniques and comparisons of the simulation result of the HO algorithm to others. The main aim was to create a speed based algorithm, though our HO decision technique also considers all the aspects of HO i.e., bandwidth, SINR (interference), traffic type, access permission as illustrated in Table-2. The preliminary simulation results show a better result than the signal strength based and speed based HO techniques and showing lower number of HO attempts in the cell centre areas. However, for future work we are analysing to improve our proposed algorithm to be energy efficient and to make it performs better in the cell edge area from other HO algorithms.

REFERENCES

[1] L. Nuaymi, WiMAX: Technology for Broadband Wireless Access. Wiley, NewYork, 2008.

[2] E Dahlman, S Parkvall, J Skold, P Beming, 3G Evolution HSPA and LTE for Mobile Broadband. Academia Press, USA, 2008.

[3] F. Capozzi, G. Piro, L. Grieco, G. Boggia, P. Camarda, "On Accurate simulations of LTE femtocells using an open source simulator," in EURASIP Journal on Wireless Communication and Networking, 2012.

[4] Onyeije Consulting LLC, Solving the capacity crunch (2011).

[5] T. Zahir, K. Arshad, A. Nakata, K. Moessner,"Interference Management in Femtocells," in Commun. Surveys & Tutorials, vol. 15, pp. 293-311, 2013.

[6] www.en.wikipedia.org/wiki/Fixed_mobile_convergence/.

[7] R. Bendlin, V. Chandrasekhar, C. Runhua, A. Ekpenyong, and E. Onggosanusi, "From homogeneous to heterogeneous networks: A 3GPP Long Term Evolution rel. 8/9 case study," in 45th Annu. Conf., Inform. Sci. and Syst. (CISS), Baltimore, MD, pp. 1-5, 2011.

[8] H. Mahmoud, I .Guvenc," A comparative study of different deployment modes for femtocell networks," in IEEE Int. Symposium on Personal, Indoor and Mobile Radio Commun. , Palo Alto, USA, pp. 1–5, 2009.

[9] D. Xenakis, N. Passas, L. Merakos, C. Verikoukis, "Mobility Management for Femtocells in LTE-Advanced: Key Aspects and Survey of Handover Decision Algorithms" in IEEE Commun. Surveys & Tutorials., vol. 16 pp. 2014.

[10] J. Moon, D. Cho, "Efficient handoff algorithm for inbound mobility in hierarchical macro/femto cell networks", IEEE Commun. Mag. Letters, vol.13, no.10, pp.755-757, Oct. 2009.

[11] P. Xu, X. Fang, R. He, Z. Xiang, "An efficient handoff algorithm based on received signal strength and wireless transmission loss in hierarchical cell networks", Telecom. Sys. J., Elsevier, pp. 1-9, Sept. 2011.

[12] D. Lopez-Perez, A. Valcarce, A. Ladanyi, G. de la Roche, J. Zhang, "Intracell handover for interference and handover mitigation in OFDMA two-tier macrocell-femtocell networks", EURASIP J. on Wirel. Com

[13] A. Ulvan, M. Ulvan, and R. Bestak, "The Enhancement of Handover Strategy by Mobility Prediction in Broadband Wirel. Access", Netw. and Electronic Commerce Research Conf. (NAEC) 2009, TX: American Telecom. Sys. Mgmt. Assoc. Inc., pp. 266-276, 2009. ISBN 978-0-9820958-2-9.

[14] A. Ulvan, R.Bestak, M.Ulvan,"Handover Scenario and Procedure in LTE-based Femtocell Networks," UBICOMM 4th Intl. Conf. on mobile ubiquitous computing, system, services and tech.Florance, Italy, 2010.

[15] H. Zhang, X. Wen, B. Wang, W. Zheng, Y. Sun, "A Novel Handover Mechanism Between Femtocell and Macrocell for LTE Based Networks", IEEE 2nd Internat. Conf. on Comm. Softw. and Nets. 2010 (ICCSN), pp.228-231, Feb. 2010.

[16] D. Xenakis, N. Passas, and C. Verikoukis, "A Novel Handover Decision Policy for Reducing Power Transmissions in the two-tier LTE network", 2012 IEEE Internat. Comm. Conf. (ICC), pp.1352-1356, June 2012.

[17] D. Xenakis, N. Passas, C. Verikoukis, "An energy-centric handover decisionalgorithm for the integrated LTE macrocell-femtocell network", Comp. Comm., Elsevier, 2012

[18] An Introduction to LTE: LTE, LTE-Advanced, SAE and 4G Mobile Communications, First Edition, m Copyright John Wiley and Sons Ltd., Inc. 2012. Published by John Wiley & Sons, Ltd, ISBN: 9781119970385

[19] G Piro, L Grieco, G Boggia, F Capozzi, P Camarda,"Simulating LTE cellular systems: an open-source framework, " in Vehicular Technology, IEEE Trans. vol. 60, pp. 498-513, 2011.

[20] T. Camp, J. Boleng, and V. Davies, "A survey of mobility models for ad hoc network research," Wireless Communications and Mobile Computing, vol. 2, pp. 483–502, Aug. 2002.

[21] R Subramanian, P Ghosal, S Barua, S Xing, S Cong, K Sandrasegaran," Survey of LTE Downlink Schedulers Algorithm in Open Access Simulation Tolls NS-3 and LTE-Sim ", Manuscript in International Journal of Wireless & Mobile Networks(IJWMN), November 2014.

Authors

Pantha Ghosal is a Graduate Research Assistant at Faculty of Engineering and Technology (FEIT), CRIN, University of Technology, Sydney. Prior to this, he completed B.Sc in Electrical and Electronic Engineering from Rajshahi University of Engineering & Technology, Bangladesh in 2007. He is an expert of Telecommunication network design and holds more than 7 years of working experience in 2G/3G and LTE. Throughout his career he was involved in several projects of RF Planning, Designing and Dimensioning.

Shouman Barua is a PhD research scholar at the University of Technology, Sydney. He received his BSc in Electrical and Electronic Engineering from Chittagong University of Engineering and Technology, Bangladesh and MSc in Information and Communication Engineering from Technische Universität Darmstadt (Technical University of Darmstadt), Germany in 2006 and 2014 respectively. He holds also more than five years extensive working experience in telecommunication sector in various roles including network planning and operation.

Ramprasad Subramanian is an experienced telecom engineer in the field of 2G/3G and LTE/LTE-A. He holds M.S (By research) in Information and Communication from Institute of Remote Sensing, Anna University (India)(2007) and Bachelors of Engineering in Electronics and Communication engineering from Bharathidasan University (2001)(India). He has done many projects in the area of 2G/3G and LTE. He has done many consultative projects across Africs/Americas/Asia etc. He was the recipient of India's best invention award for the year 2004 from Indian Institute of Management Ahmadabad and Government of India. His current research focuses on 4G mobile networks and vehicular Ad hoc networks.

Shiqi Xing is currently doing his Bachelor of Telecommunication Engineering at University of Technology, Sydney. He is an experienced programmer and currently undertaking projects in 4G Telecommunication and Robotics. Throughout his academic career he received several scholarships from The People's Republic of China.

Dr Kumbesan Sandrasegaran is an Associate Professor at UTS and Centre for Real-Time Information Networks (CRIN). He holds a PhD in Electrical Engineering from McGill University (Canada)(1994), a Master of Science Degree in Telecommunication Engineering from Essex University (1988) and a Bachelor of Science (Honours) Degree in Electrical Engineering (First Class) (1985). His current research work focuses on two main areas (a) radio resource management in mobile networks, (b) engineering of remote monitoring systems for novel applications with industry through the use of embedded systems, sensors and communications systems. He has published over 100 refereed publications and 20 consultancy reports spanning telecommunication and computing systems.

BANDWIDTH AWARE ON DEMAND MULTIPATH ROUTING IN MANETS

Tripti Sharma[1] and Dr. Vivek Kumar[2]

[1]Department of computer science & Engineering, Inderprastha Engineering College, Ghaziabad, (U.P), India.
[2]Department of computer science, Gurukul Kangri Vishwavidyalaya, Haridwar, (U.K), India.

ABSTRACT

Mobile Ad-hoc Networks (MANETs) are self configuring, decentralized and dynamic nature wireless networks which have no infrastructure. These offer a number of advantages, however the demand of high traffic flows in MANETs increases rapidly. For these demands, limited bandwidth of wireless network is the important parameter that restrains the development of real time multimedia applications. In this work, we propose a solution to utilize available bandwidth of the channel for on demand multiple disjoint paths. The approximate bandwidth of a node is used to find the available bandwidth of the path. The source chooses the primary route for data forwarding on the basis of path bandwidth. The simulation results show that the proposed solution reduces the frequency of broadcast and performs well in improving the end to end throughput, packet delivery ratio, and the end to end delay.

KEYWORDS

MANET, node-disjoint, multipath, bandwidth.

1. INTRODUCTION

Many characteristics of ad hoc networks make QoS (Quality of Service) provisioning and QoS routing, a difficult problem. QoS routing means not only to find a route from source to destination, but to find a quality route that satisfies the end-to-end QoS requirement, often given in terms of bandwidth, delay or loss probability. The single path routing protocols like DSDV and DSR, normally fail to fulfill the above requirements. The dynamic topology of MANETs provides the existence of multiple routes between two nodes, which can be utilized to transmit the packet for better support to real time communications. In case of route break, an alternate route can be used to send the packets to reduce the delay and jitter. Research shows that the use of multipath routing in ad hoc networks which are denser performs better throughput. In this paper, we are proposing a multipath routing protocol, which is the potential improvement of the existing Ad hoc On-demand Multipath Distance Vector (AOMDV) protocol [1], and that could be achieved when utilizing the bandwidth of the channel and bandwidth of the respective paths.

The remaining part of the paper is organized as follows: the next part is the review of the protocols and methodologies in the required fields of MANET. Then, we present the problems and motivations. Since, we have modified the existing AOMDV routing protocol, it is also discussed briefly, with its problems. And then we propose the improvements in AOMDV followed by the simulation results of the comparison of new protocol with the existing one. In the last section, we provide the conclusion and future scope of the proposed work.

2. RELATED WORK

In [3-13], various approaches to QoS provisioning and QoS routing for single path in MANETs have been studied and derived with the aim to reduce the connection set up latency, delay and bandwidth and to ensure guaranteed performance level to the QoS sensitive applications. Multipath routing is more promising in ad hoc networks since it provides additional features like load balancing, fault-tolerance, higher throughput etc., to ensure QoS assurance in ad hoc networks.

Zhi Zhang, et al., [14] performs bandwidth estimation method with the on-demand node-disjoint multi-path routing protocol. This approach creates the multiple node-disjoint paths during the route discovery process and maintained those paths actively. The detector packets measure the available bandwidth of each hop along the paths. AOMDV uses the basic idea of the popular ad hoc on-demand distance vector (AODV) which is proposed in [2]. AOMDV extends the AODV protocol to find the multiple paths in the route discovery process without discarding those paths. These all multiple paths are guaranteed to be loop-free and disjoint. AOMDV has three important issues compared to other on-demand multipath routing schemes. Firstly, like some other protocols (e.g., TORA, ROAM [15-16]), have high coordination overhead among nodes, but its inter-nodal coordination overheads is less. Second, the disjointness of alternate routes is guaranteed via distributed computation without the use of source routing. Finally, this protocol computes alternate paths over AODV with very less additional overhead; it does this by utilizing the information which is already available with alternate paths as much as possible. There are a number of extensions of AOMDV in various fields including dynamic route switching, stability, load-balancing and randomization.

As the AOMDV is based on static route selection, it could not handle the change of the network such as congestion and contention. D. Shin et al., [17], proposed A2OMDV (Adaptive AOMDV), in this approach author resolve the problem of AOMDV, through dynamic route switching method. A source node finds its route dynamically based on the delay of the multiple paths and observes the quality of the alternative routes according to the change of the ad hoc network. One idea is to accept partially disjoint paths that are more stable than other maximally disjoint paths that could increase paths lifetime. Stability-based Partially Disjoint AOMDV (SPDA) protocol is proposed in [18], which is a modification of the AOMDV protocol, finds partially disjoint paths based on links stability. These Partially Disjoint paths improves MANET performance in terms of delay, routing packets overhead, and the network throughput. M. Tekaya, et al., [19], also, introduced a multipath routing protocol with load balancing mechanism, to develop a new protocol called QLB-AOMDV (QoS and Load Balancing- Ad Hoc On demand Multipath Distance Vector), with this solution we can achieve better load balancing with respect to the end-to-end QoS requirement. A multipath routing algorithm is proposed by Pinesh A Darji, et al. in [20], that could randomize delivery paths for data transmission also it uses secured traffic load based on some cryptography approach, in which, randomized paths can protect data from the intrusion of malicious nodes. In [21], author proposed an adaptive retransmission limits algorithm for IEEE 802.11 MAC to reduce the false link failures and predict the node mobility. Since the probability that neighbour node is still in transmission range and may be not responding due to some problems other then mobility is maximum. In this approach the signal strength of each node in network is monitored and, while performing transmissions to a neighbour node, if it's received signal strength is raised and is received recently then Adaptive MAC persists in its retransmission attempts.

2.1 Overview of AOMDV

Ad-hoc On-demand Multi path Distance Vector Routing (AOMDV) protocol is an extension to the AODV protocol for computing multiple loop-free and link disjoint paths. The routing entries for each destination contain a list of the next-hops along with the corresponding hop counts. Multiple paths maintained at each node for each destination have the same destination sequence number which helps in keeping track of the route. For each destination, a node maintains the advertised hop count, which is defined as the maximum hop count for all the paths, which is used for sending route advertisements of the destination. The duplicate route advertisement received by a node defines an alternate path to the same destination. Loop freedom is assured for a node by accepting alternate paths to destination if it has a less hop count than the advertised hop count for that destination. Because the maximum hop count is used, the advertised hop count therefore does not change for the same sequence number. When a route advertisement is received for a destination with a greater sequence number, the next-hop list and the advertised hop count are reinitialized.

AOMDV can be used to find node-disjoint or link-disjoint routes. To find node-disjoint routes, each node does not immediately reject duplicate RREQs. Each RREQs arriving via a different neighbor of the source defines a node-disjoint path. Since the nodes cannot broadcast duplicate RREQs, so any two RREQs arriving at an intermediate node through a different neighbor of the source could not have traversed the same node. In an attempt to get multiple link-disjoint routes, the destination replies to duplicate RREQs, the destination only replies to RREQs arriving via unique neighbors. After the first hop, the RREPs follow the reverse paths, which are node disjoint and thus link-disjoint. The trajectories of each RREP may intersect at an intermediate node, but each takes a different reverse path to the source to ensure link disjointness. Using AOMDV protocol is advantageous since it allows intermediate nodes to reply to RREQs, while still selecting disjoint paths. One of the drawbacks of AOMDV is, it has more message overheads during route discovery due to increased flooding and since it is a multipath routing protocol, the destination replies to the multiple RREQs that results are in larger overhead.

2.2 Problems And Motivations

It is concluded from above discussion, that a reactive routing protocol generates a large number of overhead control messages in the network during route discovery process. So, routing with QoS is difficult in MANET due to several reasons like high overhead, dynamic nature of MANETs, and guarantee of reserved resources.

The existing AOMDV protocol has given the improved results compared to AODV. As it does not have large inter-nodal coordination overheads, it provides disjoint alternate routes, and these are with minimal additional overhead over AODV. Still, there are some problems in AOMDV which are considered in various modifications of it. Those modifications have been discussed in section 2. The modified protocols have resolved the problems of AOMDV up to some extent, but none of them has considered the resource utilization of the MANET. There are various problems in AOMDV extensions, these are:

- None of the protocol has given the strategy to prioritize the alternate routes for resource utilization.
- The selection of the alternate routes performed without comparison of performances.
- The existing protocols are not effective to compute more disjoint paths between source and destination pairs while considering the effect of other resources on performance of the network.

3. THE PROPOSED PROTOCOL

In this section, we propose an extension to AOMDV protocol, in which the channel bandwidth is utilizes in order to improve the network performance. AOMDV allows finding many routes between source and destination during the same route discovery procedure that guarantees loop freedom and disjointness of alternate paths, and only one path is used to transmit data. The routing table entry structure of AOMDV is modified for the proposed method in which only one field is added which gives the information about the path bandwidth of the multiple paths stored in route list entries and is given in Figure 1.

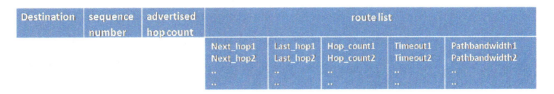

Figure 1. Routing table entry structure of proposed protocol

3.1 Route discovery

The route discovery procedure of the proposed method constructs multiple bandwidth aware paths between a source and destination. Route request and Route reply packets now contain the existing information and the available bandwidth of the node forwarding it. The source is able to learn the bandwidth of the multiple paths during the route discovery by using the Maximum-Minimum approach to measure the quality of the path. In this approach, the available bandwidth of the entire path is just the available bandwidth of the weakest link. Once the source receives the RREPs, it stores its next hop information and chooses the path with the greatest available bandwidth as its primary path for data transmission. The bandwidth of the route is determined periodically in order to find the optimal route in the change of the network topology with the help of *detector* packet as explained in section 3.4. The source node will switch from its current primary path to an alternate path if the difference in their available bandwidth is higher than the predefined threshold in contrast to wait for its primary path to break. In Figure 2, the AOMDV route decision procedure is summarized, which is modified for the proposed solution, and is given in Figure 3.

```
If (no route to destination)
{
Initiate route discovery as in AODV;
}
If (single known route)
{
Forward data packet to specified route;
}
Else //if N routes are known from source to destination.
{
Forward data packet to best route;
// on the basis of minimum hop count.
}
```

Figure 2. AOMDV route decision

```
If (no route to destination)
{
Initiate route discovery as in AOMDV;
}
If (single known route)
{
Forward data packet to specified route;
}
Else
// if N routes are known from source to destination
{
Forward data packet to route with max. available bandwidth ;
}
```

Figure 3. Modified AOMDV route decision

We also propose to modify the route request as well as route reply packet, which are given below:

$RREQ(proposed) = RREQ(AOMDV) + ABW (node)$
where,
$ABW (node) = MIN [ABW (RREQ\ recieved), B_{avail} (node)]$

In the proposed RREQ packet, the B_{avail} of node is the available bandwidth of the node sending the packet. The estimation of B_{avail} is discussed in section 3.3. Upon receiving the packet, each node will compare its own available bandwidth with the bandwidth received in the packet, and then update the packet with the minimum bandwidth. Once the destination receives the RREQ, it generates route reply packet ie, RREP which is also modified in the similar manner.

3.2 Hello message

The Hello packet used in AOMDV only keeps the address of the node which has generates this packet. We modify the Hello packet for the new solution by adding the bandwidth information of the node sending the hello packet and the neighbors of the node with their bandwidth information. Each node broadcast this hello packet periodically, and updates all its neighbors about its bandwidth. The format of hello packet is given in Figure 4. Where $B_{consumed}$ is the bandwidth consumed by each node for sending packets in the network.

| <SenderID, $B_{consumed}$, timestamp> | <neighborsID, $B_{consumed}$, timestamp> |

Figure 4. Format of Hello message

3.3 Bandwidth Estimation

For the forwarding of data packets where multiple routes are known from source to destination, the maximum available bandwidth of the routes is estimated. Each node estimates its consumed bandwidth by tracking the packets it transmits into the network. This value is recorded in the bandwidth consumption register at the node and updated periodically. Once a node knows the bandwidth consumption of its one-hop neighbors and its two-hop neighbors, the residual bandwidth can be estimated as (1), the raw channel bandwidth (B_{raw}) minus the overall consumed bandwidth ($B_{all\_consumed}$), multiplied by a weight factor. We need to multiply the residual bandwidth by a weight factor α due to overhead of IEEE 802.11 MAC, overhead of routing protocol and overhead for the situation where a node is in sender's interference range but it isn't in any of sender's neighbors' transmission range [14]. In this situation, the sender will never know this node bandwidth usage. However, these instances do not happen frequently since it has to meet strict requirements. So weight factor is used to overcome this situation. From the equation in (1), the more interference traffic in the channel the more conservative the estimation will be.

$$B_{avail} = \alpha\,(B_{raw} - B_{all\_consumed}) \qquad (1)$$
$$\text{where, } 0 < \alpha < 0.8865$$

3.4 Alternate Route Maintenance

The alternate routes constructed between a pair of source and destination is to be maintained for a time period. The algorithm for alternate route maintenance is given below. The *detector* packet is unicast from source to destination along the alternate paths. This packet contains one field apart from source and destination addresses that is to collect the minimum bandwidth along each path.

Algorithm :

1. Source node periodically sends detector packet to the destination along each of its alternate paths after route discovery.
2. Each node updates the bandwidth field when the detector propagates through the alternate paths.
3. The destination records the bandwidth in the detector and sends a new detector back to the source along the same path.
4. The bandwidth of the entire path is just the bandwidth of the weakest link.
5. The source node chooses the path with the maximum bandwidth for routing.
6. The source node will switch from its current primary path to an alternate path if the difference in their ABW is higher than the predefined threshold in contrast to waits for its primary path to break.

4. SIMULATION RESULTS

We study the new AOMDV performance using ns-2 [22, 23] simulations. The main objective of our simulation is to evaluate the effectiveness of new AOMDV relative to AOMDV in the presence of mobility-related route failures. Other objective includes evaluating the number of alternate node disjoint paths that can be found using new AOMDV.

4.1 Simulation Environment

The simulation experiment is carried out in LINUX (ubuntu 10.4). The detailed simulation model based on network simulator-2 (ver-2.35), is used in the evaluation. Table 1 shows the simulation parameters. In this simulation, each packet starts its journey from a random location to a random destination with a randomly chosen speed (uniformly distributed between 0–20 m/s). Simulations are run for 50s, 100s, 150s, 200s, 250s and 300s simulated for 100 nodes under CBR traffic pattern. The weight factor α is defined as 0.65.

Table 1. Simulation Parameters

Parameter	Values
Dimensions	1000m×1000m
Traffic type	CBR
Number of nodes	100
Simulation Time	300s

Pause Time	50, 100, 150, 200, 250, 300s
Total Sources and	49 and 71
Maximum Speed of Nodes	20m/s
Packet rate	4pkts/s
Packet size	512 byte
Mobility model	Random Waypoint Model
Channel bandwidth	2Mbps

4.2 Performance Metrics

The performance of routing protocol is evaluated using three different metrics to compare the performance of the new protocol with the existing AOMDV routing protocol. They are:

1. **Packet delivery fraction (PDF)** — The packet delivery fraction is the ratio of the data packets delivered to the destinations to those generated by the sources.

2. **End-to-end delay (E2E Delay)** — The end-to-end delay of data packets refers to the time taken for a packet to be transmitted across a network from source to destination.

3. **Normalized routing load (NRL)** — The Normalized routing loads is computed by the ratio of total number of routing packets sent by the number of data packets delivered successfully.

4. **Throughput (THPT)** — The throughput is the amount of data packets received at the destination per unit time.

The performance results of AOMDV and B-AOMDV for 100 nodes and the comparison of new protocol with the existing AOMDV protocol are given below in Table 2 and 3.

Table 2. Performance Results of AOMDV and B-AOMDV for 100 nodes

Simulation time (s)	B-AOMDV				AOMDV			
	PDF (%)	E2E Delay (s)	NRL	THPT	PDF (%)	E2E Delay (s)	NRL	THPT
50	90.95	0.01301	2.636	653.11	93.33	0.022	2.908	587
100	92.43	0.09289	1.502	958.33	95.206	0.0208	1.587	1027
150	79.52	1.82	1.261	1300	75.25	1.35	1.95	1295
200	64.27	3.67	1.21	1516.35	59	3.35	3.403	1490
250	56.063	5.26	1.258	1633.09	50	4.86	4.49	1629.3
300	51.101	5.99	1.402	1706.63	42	5.96	4.96	1711.6

Table 3. Comparison Table

Metrics	B-AOMDV	AOMDV
PDF	High	Low
E2E	Low	High
NRL	Low	High
THPT	High	Low

Packet Delivery Fraction: Figure 5 compares the packet delivery fraction of AOMDV and proposed modification in varying pause time and random node speed. The graph demonstrates that proposed modification performs better than the AOMDV at nearly all pauses of time. The AOMDV perform well at less pause time but degrade at high pause time, while the proposed protocol does not degrade too much. Higher packet delivery fraction of new protocol is because of the availability of the bandwidth utilization among alternate paths to forward the packets when the source switched from its primary path.

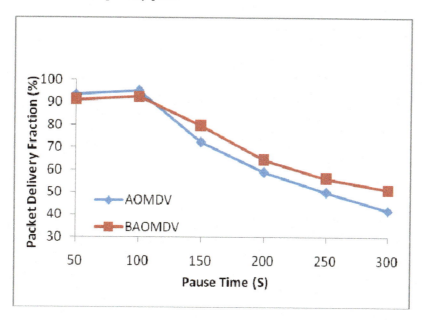

Figure 5. Packet delivery fraction

End to end delay: Figure 6 compares the End to end delay of AOMDV and proposed modification in varying pause time and random node speed. The graph demonstrates that proposed modification results in less delay than the AOMDV at nearly all pause time. The AOMDV perform well at less pause time but delay increase at high pause time, while the proposed protocol does not increase the delay at almost all pause time.

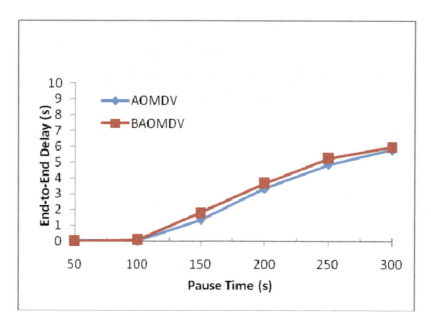

Figure 6. End-to-End Delay

Normalized Routing Load: Figure 7 compares the Normalized Routing Load of AOMDV and proposed modification in varying pause time and random node speed. The graph demonstrates that proposed modification results in less normalized routing load than the AOMDV. The AOMDV perform well at less pause time but load increase at high pause time, while the proposed protocol results high normalized load at less pause time but normalized routing load is decreased as the pause tme is increased.

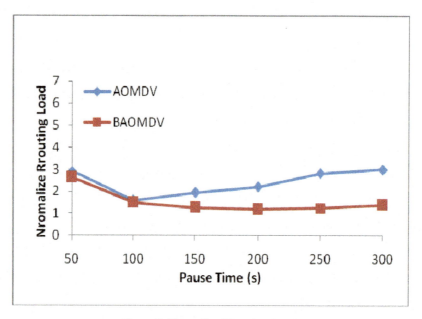

Figure 7. Normalized Routing Load

Throughput : Figure 8 compares the throughput of AOMDV and proposed modification in varying pause time and random node speed. The graph demonstrates that proposed modification results in high throughput than the AOMDV. The AOMDV performs well at less pause time but at high pause time, the new protocol results high throughput as compared to AOMDV protocol.

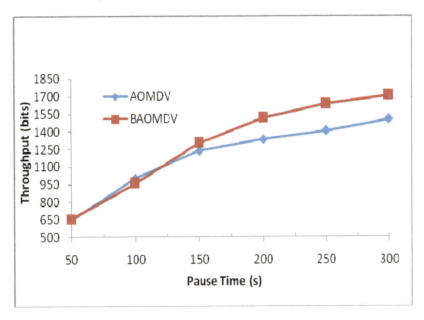

Figure 8. Throughput

5 CONCLUSION AND FUTURE SCOPE:

In this paper, we proposed an approach for multi-path routing in mobile ad hoc networks and used bandwidth estimation by disseminating bandwidth information through *detector* packets. The primary property of this approach is that it can adapt the change in network topology by proactively estimating the available bandwidth of each path to the destination and always using the best path. Simulation results show that the performance of the protocol is superior to the AOMDV in all most all scenarios. Future research will focus on optimally distributing traffic over multiple paths to upgrade the performance of the protocol.

REFERENCES

[1] Marina, M. K., and Das, S.R. (2006) "Ad hoc on-demand multipath distance vector routing", Wireless Communications and Mobile Computing, vol. 6, no. 7, pp 969–988,.

[2] Perkins, C.E., Royer, E. (1999) "Ad-hoc On Demand Distance Vector Routing", IEEE Workshop on Mobile Computing Systems and Applications (WMCSA'99), pp 90-100,.

[3] Chakrabarti, S., Mishra, A. (2001) "QoS issues in Ad Hoc Wireless Networks", IEEE Communication Magazine, feb, pp 142-148,

[4] Lohier, S. et.al. (2003) "A reactive QoS Routing Protocol for Ad Hoc Networks", European Symposium on Ambient Intelligence (EUSAI'2003), LNCS, Nov, Springer Verlag, Netherlands, 875, pp 27-41.

[5] Cerd`a, L., et. al. (2005) "A Reservation Scheme Satisfying Bandwidth QoS Constraints for Multirate Ad-hoc Networks", Eurasip.org, conference proceedings, IST, jun.

[6] Chen, L. and Heinzelman, W. B. (2005) "QoS-Aware Routing Based on Bandwidth Estimation for Mobile Ad Hoc Networks", IEEE Journal On Selected Areas In Communications, vol. 23, no. 3, pp 561-572.

[7] Geng , R. and Li, Z. (2006) "QoS-Aware Routing Based on Local Information for Mobile Ad Hoc Networks", S. Madria et al. (Eds.): ICDCIT, LNCS 4317 © Springer-Verlag Berlin Heidelberg, pp 159 – 167.

[8] Chen, L. and Heinzelman, W. B. (2007) "A Survey of Routing Protocols that Support QoS in Mobile Ad Hoc Networks", IEEE Network, vol. 21, no. 6, pp 30-38.
[9] Marwaha, S., Indulska, J. and Portmann, M. (2008) "Challenges and Recent Advances in QoS Provisioning, Signaling, Routing and MAC protocols for MANETs", IEEE conference on Telecommunication Networks and Applications, pp 97-108.
[10] Sharma, N. et.al, (2010) "Provisioning of Quality of Service in MANETs Performance Analysis & Comparison (AODV and DSR)", IEEE conference on Computer Engineering and Technology, vol. 7, pp 243-248.
[11] Shenoy, N., Pan, Y. and Reddy, V.G. (2005) "Quality of Service in Internet MANETs", IEEE 16th International Symposium on Personal, Indoor and Mobile Radio Communications, pp 1823-1829.
[12] Medhi, D. (2002) "Quality of Service (QoS) Routing Computation with Path Caching: A Framework and Network Performance", IEEE Communications Magazine, vol. 40, no. 12, pp 106–113.
[13] Medhi, D. and Sukiman, I. (2000) "Multi-Service Dynamic QoS Routing Schemes with Call Admission Control: A Comparative Study", IEEE Journal of Network and Systems Management, vol. 8, no.2, pp 157–190.
[14] Zhang, Z., Dai, G. and Mu, D. (2006) "Bandwidth-Aware Multipath Routing Protocol for Mobile Ad Hoc Networks", J. Ma et al. (Eds.): UIC LNCS 4159 © Springer-Verlag Berlin Heidelberg, pp 322 – 330.
[15] Park, V.D. and Corson, M.S. (1997) "A highly adaptive distributed routing algorithm for mobile wireless networks", Annual Joint Conference of the IEEE Computer and Communications Societies. Proceedings IEEE, vol. 3, pp 1405–1413.
[16] Raju, J. and Garcia-Luna-Aceves, J. (1999) "A new approach to on-demand loop-free multipath routing", Proceedings of IEEE International Conference on Computer Communications and Networks, pp 522 – 527.
[17] Shin, D., et. al. (2009) "A2OMDV : An Adaptive Ad Hoc On-Demand Multipath Distance Vector Routing Protocol Using Dynamic Route Switching", Journal of Engineering Science and Technology, vol. 4, no. 2, pp 171 – 183.
[18] AlMobaideen, W. (2009) "SPDA: Stability Based Partially Disjoint AOMDV", European Journal of Scientific Research, vol. 27, no. 3, pp 342-348.
[19] Tekaya, M., Tabbane, N. and Tabbane, S. (2010) "Multipath Routing with Load Balancing and QoS in Ad hoc Network", International Journal of Computer Science and Network Security, vol. 10, no. 8, pp 280-286.
[20] Darji, Pinesh. A. et. al., (2012) "An Improvement in AOMDV with Randomization", International Journal of Computer Technology & Applications, vol. 3, no. 3, pp 968-972.
[21] Bhatia, G. and Kumar, V., (2010) "Adapting MAC 802.11for performance optimization of MANET Using Cross Layer Interaction", International Journal of Wireless & Mobile Networks (IJWMN) Vol.2, No.4, pp 31-42.
[21] McCanne, S. and Floyd, S. ns Network Simulator, http: //www. isi.edu/nsnam/ ns.
[22] Fall, K. and varadhan, k. the VINT project. The NS manual.

HIGHLY RELIABLE MULTI-SERVICE PROVISIONING USING SEQUENTIAL PREDICTION OF ZONE AND PL&T OF NODES IN MOBILE NETWORKS

Sharmistha Khan[1], Dr. Dhadesugoor R. Vaman[2], Siew T. Koay[3]

[1] Doctoral Student, Electrical and Computer Engineering Department and ARO CeBCom,
Prairie View A&M University, Prairie View, TX, USA
[2] Texas A&M University System Regents Professor and Director of ARO CeBCom,
Electrical and Computer Engineering Department, Prairie View A&M
University, Prairie View, TX, USA
[3] Professor, Electrical and Computer Engineering Department, Prairie View A&M
University, Prairie View, TX, USA

ABSTRACT

In this paper, a dynamic Position Location and Tracking (PL&T) method is proposed which uses an integrated "zone finding by predictive algorithm" and "Triangulation with dynamic reference positioning" for target nodes. The nodes can be used as references or targets at different instances of time to maintain accurate PL&T of each node in a wireless mobile network. Each node is equipped with an omnidirectional antenna. It is shown that while "zone prediction" is fairly accurate for finding the PL&T of a target node based on previous good location data when the target is moving linearly in one direction, it is not adequate when the nodes change directions. However, once the zone is predicted, the triangulation will provide accurate position of the target node by placing the reference nodes to form a reasonably perfect geometry. The results show that triangulation alone does not provide accurate prediction of the target location without finding the zone of the target and zone finding alone does not support accuracy of PL&T when the target node is changing directions. Another aspect that should be of concern is the repeated PL&T operation on the same target node at different instance of time becomes erroneous and therefore, they need to be reinitialized to get accurate locations with known wait points. Also, the references need to be reinitialized to maintain the accuracy. This requires the nodes to be dynamically used as target nodes and with good PL&T, they are used as references at different time instances. There is an upper layer dynamic reference management is required for this technology. The proposed integrated predictive zone finding and triangulation is particularly suitable for clustered Mobile Ad hoc Networks (MANET). Finally, this paper shows the performance analysis of this system for a reasonable number of nodes within the cluster that demonstrates high accuracy of PL&T location data of the nodes. Of course the overhead in managing the dynamic references is increased. However, the frequency of changing the references is minimized in each cluster as it contains few nodes (40 or less).

KEYWORDS

Position, Localization, Tracking, Zone Prediction, Triangulation

1. INTRODUCTION

Currently, a variety of wireless technologies are rapidly emerging for supporting multi-service provisioning. These technologies include: Wireless Metropolitan Area Networks (WMAN), Wireless Local Area Networks (WLAN), Wireless Wide Area Networks (WWAN), the Worldwide Interoperability for Microwave Access (WiMAX), General Packet Radio Service

(GPRS), Universal Mobile Telecommunications System (UMTS), and Wireless Fidelity (WiFi). They are being deployed as part of the overall Internet Protocol (IP) based infrastructure and are standardized to provide high speed data rate, high-quality, and high capacity multimedia services in different geographic areas. They also support mobility whereby users can access the network centric applications wherever they have access to these networks and not limited to their home base. In addition, the deployment of backhaul wireless networks is facilitating the users to have greater access broadband multi-services applications. However, in mobile communication uninterrupted network connectivity is still challenging. It is really essential to provide continuous network connectivity to satisfy high levels of mobile service quality and maintain the prescribed levels of security and privacy. The most challenge today in wireless network research is to support end-to-end quality of service to a wide variety of multi-services that includes both real time and non-real time applications. To provide continuous network connectivity, it is very essential to get information about the required transmission power for maintaining the connectivity between a source and a destination and their location identification to support better management for connectivity. The Position, Location and Tracking (PL&T) of a node needs to be accurately maintained at all time. Target or object (or node) PL&T has become critical in many applications such as navigation, tracking, emergency service, location based services (LBS) and security. Several algorithms have been already built for tracking the target node accurately. Most of them have used the received signal power methodology. However it is really difficult to find out the location of a target accurately as the received signal tends to be distorted due to jitter, channel noise, and interference. Therefore, still some improvement and innovation is required to achieve the PL&T accuracy.

In this paper, we propose a PL&T algorithm based on a zone prediction and triangulation that allows the nodes to use omnidirectional antennas. Tracking accuracy needs to be maintained to ensure power efficiency and connectivity at all times. Currently most of the commercial systems use Global Positioning System (GPS) for node PL&T while it is moving. However, GPS algorithms tend to be inaccurate near buildings and when the nodes are not visible due to weather or blocked areas. GPS also cannot be used indoors to predict PL&T as it does not distinguish different floor [1]. In this paper, we propose to use nodes which have accurate PL&T at an instance of time as references for triangulation which is applied after we predict the zone of the target node. The zone prediction prior to triangulation allows the reference nodes to be places to form a good geometry for triangulation to improve the PL&T accuracy. Our proposed algorithm will able to track any moving objects as well as stationary objects. Our algorithm does not restrict the directional movement of the target nodes. The performance of the combined zone prediction and triangulation will be demonstrated for its accuracy.

In zone prediction algorithm, the target location at an instance of time is predicted based on previous good locations using an iterative process. Once the zone of the target is predicted with respect a relative origin, the nodes that have accurate PL&T will be used as reference nodes to form a geometry and triangulation will be performed to accurately predict the PL&T of the target node at that instance of time. As the target node keeps moving (in any direction), the reference nodes will be dynamically changed to maintain proper geometry for computation of PL&T of the target node at each instance of time after the zone is predicted.

2. BACKGROUND

The demand of uninterrupted services is rapidly increasing with the current growth of smart mobile devices, multimedia services such as audio- video streaming, and different location based services .To provide the uninterrupted network services, Position, Location and Tracking (PL&T) of mobile nodes in a network has become a very important methodology as an alternate method of using power measurements. Average power measurements are impacted by the interference such as multi-fading, whereas PL&T of nodes with good measurements is more reasonable in assessing the location of a node with respect to a sending node in order to maintain power efficiency and also connectivity between nodes. The use of PL&T technique has become very useful in both homeland and battlefield theaters to avoid friendly casualties. As long as the accuracy of PL&T of nodes is maintained within reasonable error, this technique is also very useful in applications where fire hazards may occur and saving life within a constrained time is very critical. Finally, it is possible to see that PL&T technique has been used in numerous applications today including WSN, navigation, tracking, indoor positioning, emergency 911 in USA, and emergency 112 in Europe.

Researchers have developed various PL&T algorithms as can be seen in published literature. However, accuracy of these algorithms in tracking moving nodes is still a challenge in many applications. Until, one can develop very accurate predictions using PL&T, their applications for mission critical areas will be difficult.

To identify the PL&T of a target, different PL&T techniques have been developed using friendly reference nodes: single reference node based PL&T [2], two reference nodes based PL&T [3], three reference nodes based Triangulation [4], and four reference nodes based multilateration methods and multiple targets' PL&T [4] where each of above methods is analysed by developing theoretical models for PL&T precision. Some of the important PL&T methods are reviewed through literature are described below.

Forward Movement based Prediction: We have found that in this method, only zone prediction method is focused within a constrained forward movement of a target node. It is based on single reference node that considers limited random movement and does not consider sharp turns or obstacles [5].

Cooperative Indoor Position Location technique: This is based on Parallel Projection Method and designed for collaborative position location to effectively handle Non Line Of Sight (NLOS) propagation based on set theory (specifically on the parallel projection method). This method is not applicable for mobile node as well as for MIMO technology. More time consuming as it is based on set theory; it will take additional time to achieve satisfactory level of accuracy [6].

Received Signal Strength based Prediction: This method provides better tracking accuracy over the triangulation method as long as multi-path fading is small and longer averaging availability. However, without zone finding, the signal is spread and errors are accumulated in the computation of both the distance and the angle for limited non-random trajectory [7].

Multi hop based Prediction: All the position locations of nodes are estimated using multiple levels of reference nodes which Increases cumulative errors in multi hop measurements and has no beam adaptation used. Thus, significantly reduces accuracy of tracking [8].

Directional Lines Intersection based Prediction: Localization of nodes using point of intersection of highly directional beams has better accuracy for low speed mobile anchor nodes. In this method, higher overhead in scanning and does not address random trajectory [9].

Distributed Position Localization and Tracking (DPLT) Method: It is a robust, reliable, and low complexity method of Distributed PL&T detection of malicious nodes in a Cluster based MANET is proposed [4]. However, only forward movement is considered for determining the adaptive beam formation.

By considering the major limitations of all the above methods, we have proposed this novel PL&T method that uses IP based triangulation based on zone prediction method where accuracies of less than a meter are required. However, multi-path fading and noise handling was a serious issue. In this method, nodes require re-initializing after repeated tracking at different instances of time to maintain the tracking error within the specified target. The reference nodes also are changed dynamically as the target nodes move. Therefore, inter-changing the reference nodes and target nodes was dynamically accomplished to maintain accuracy. Thus, this method may increase the overall complexity. On the other hand, multi-path fading and Doppler effects play a significant role in reducing the accuracies of the PL&T. Many researchers are addressing these issues through equalization, estimation and coding techniques [10-11]. Some researchers use directional antenna which basically allows concentration of beam power in one direction in order to increase the signal strength and increase the probability of handling interferences and improve the PL&T accuracy [12]. Researchers have also used Directional Antennas in interacting of steered or switched antenna systems in an ad hoc network [13]. However, the main limitation on using the directional antenna is the incapability on using for Multiple Input Multiple Output (MIMO) technology. Again, it is very difficult to use directional antenna in the end user's device. In recent years, in wireless communication field, for getting satisfactory performances, significant progress has been made in developing the overall systems. One of the most beneficial is the use of multiple antennas at the transmitter and at the receiver [14] which known as multiple input multiple output (MIMO) systems. Over using the single input single output (SISO) system, there are mainly two types of benefits on using MIMO system as well as on using multiple antesnnas: spatial diversity improvement and throughput [15]. Therefore, omnidirectional antenna is really beneficial for mobile and wireless devices those have MIMO technology as it can provide a 360 degree horizontal radiation pattern [16]. Considering this issue, we have used omnidirectional antenna in this research.

3. SYSTEM ARCHITECTURE

The proposed design of PL&T system uses a sequential process of "zone finding" followed by placing references to locate the node with a reasonable geometry that minimizes the error in triangulation with minimal outliers. In this design, we quantize a time such that the computational time for integrated zone finding and triangulation of a specific location is significantly smaller and allows the node to be virtually stationary for computation. The next location will be identified at a periodic quantized time. This process does not take into account the location in between two quantized times. It is anticipated that the importance of quantizing the time will allow computational accuracy and at the same time the transient period is not critical for the overall tracking of the mobile node. PL&T is a continuous operation of finding the location of the node with successive zone finding, placing the references for proper triangulation geometry and performing the triangulation in an atomic operation. Our proposed algorithm will able to find out the location of a target for any direction continuously since it can track the location based on both zone prediction and triangulation method. If a target node moves linearly, through zone prediction method we can predict the location accurately. However, on the other side, when a target does not move linearly means changes it's direction such as though spiral way, trapezoidal way, it required triangulation method with zone prediction to track the location accurately. The proposed algorithm that involves several steps with design architecture is described through Figure 1. Figure 1 shows the basic design architecture of our PL&T system for a particular scenario where we consider target node moves linearly.

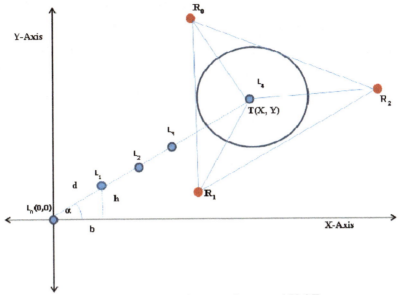

Figure 1. System Architecture of proposed PL&T

The scenario consist of four mobile nodes where R_0, R_1, R_2 are the three reference node and $T(X, Y)$ is the target node. L_0, L_1, L_2, L_3 shows the first four location point of the target node and the circle around the target node whose radius is the half distance between third and fourth point shows the predicted zone of the target node's location point . Based on previous four locations point, we have to find out the current location of a target continuously. Each time we have to find out the predicted zone of the target node's location and after finding the zone, we have to place three reference nodes (R_0, R_1, R_2) that form three triangles $\Delta R_0 T R_2$, $\Delta R_0 T R_1$, $\Delta R_1 T R_2$ with the target node for performing triangulation process.

3.1 Zone Finding

In this research, Zone finding is one of the important processes that basically predict the zone of a target's location of a particular instance of time. To implement this process we need to do the following steps that are described below in details.

3.1.1 Develop Zone Finder (ZF) algorithm

Basically, ZF algorithm is an iterative process that helps to find out some previous good location of a target node with respect of an origin. Basic system configuration of ZF algorithm is presented by Figure 2 which mainly shows the relationship between different elements of this system and their connectivity with each other and consists of a recursive polynomial executor process, mapping and correction algorithm, transport function and database for reporting data. For identifying the next location based on previous four locations, a recursive polynomial executor is used that is shown in Figure 2. IP time tag approach is used by geometric pairing to measures the distances and corrects to match the object location to a common vertex in two or more triangles. The number of triangles those are used in this algorithm is directly proportional to the number of reference points used. The minimum requirement of reference points is two for making a geometric pairing triangulation. For improving the accuracy of tracking objects, the first level of adjustment is provided by this geometric pairing triangulation. For getting high accuracy (>> 1 meter) in object tracking, an independent second level verification is provided by the ZF polynomial.

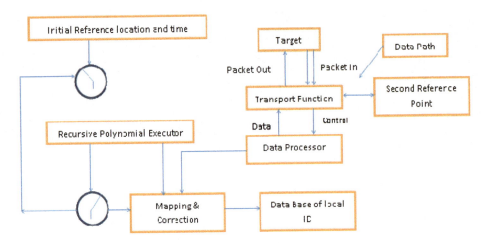

Figure 1. ZF Reference System Configuration

Furthermore, the distances and maps are used to coordinates that are consistent with the GPS coordinates by the reference system configuration. By using a basic IP transport system, within any network centric environment this data can be used for transporting purposes. IP transport is the basic transport method for exchanging data. An IP address is uses as the ID of the object and that can be changed to any customer combat ID.

3.1.2 Draw the Circle

By implementing the ZF algorithm, we can get the distance between third and fourth location point. We have considered the fourth location point as the target location. To find out the co-ordinate of the fourth location point we have to assume some measurement that is shown in Figure 1.

We can assume,

- The co-ordinate of first location point is L_0 (0, 0).
- Let, the first location point makes angel θ with the X Axis through its true path.
- From ZF algorithm (We have discussed above), we have the trajectory locations (the distance between one point to another point) of a node for an instance of time. As we have the distances between two consecutive location points (1st location point to 2nd location point and 2nd to 3rd location point), by applying the trigonometric formula, we can get the co-ordinate of the 2nd point, 3rd point, and 4th point accordingly.

We can find out the value of X-axis and Y-axis by the following equations where b is the coordinate value of X-axis, h is the coordinate value of Y-axis and d is the distance between two consecutive location points.

$$\cos\theta = \frac{h}{d} \ldots\ldots\ldots (3.1)$$
$$b = d\cos\theta \ldots\ldots\ldots (3.2)$$
$$\sin\theta = \frac{h}{d} \ldots\ldots\ldots (3.3)$$
$$h = d\sin\theta \ldots\ldots\ldots (3.4)$$

After getting the co-ordinate of the 4th location point, we have to draw a circle around the 4th point with a particular radius where the value of the radius will be the half distance between 3rd and 4th point. Basically, we will consider the perimeter of this circle as a zone for the target.

3.1.3 Place Reference Point

We have to place three reference points those have the accurate PL&T such as $R_0(X_0,Y_0)$, $R_1(X_1,Y_1)$, $R_2(X_2,Y_2)$ around the target node such a way that they will be outside of the zone and will form three triangles with good geometry.

3.2 Triangulation

Triangulation method is one of most popular positioning location techniques in which geometric properties of triangles are used to estimate a node positioning. Basic principle of this method based on three points including two reference node points and a target node point. Basically the triangles will be formed between any two reference nodes and the target node. Triangulation based PL&T has drawback of the cumulative error during localization and tracking as nodes are continuously tracked at different locations [3]. To get good geometry, we have to carefully place the reference nodes such that there is no angle less than 30 degrees or greater than 150 degrees on the triangles.

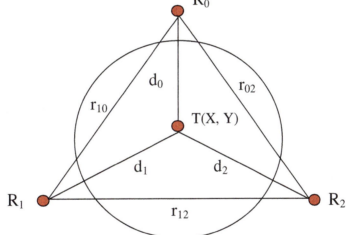

Figure 3. Triangulation with three reference nodes around the predicted zone

After forming the triangles with good geometry we have to determine the range between each reference node and the target node. Figure 3 shows the design of the triangulation with three reference nodes around the predicted zone where R_0, R_1, R_2 are three reference nodes and T is the target node that form three triangles; $\Delta R_0 TR_2$, $\Delta R_0 TR_2$, and $\Delta R_0 TR_2$. The ranges between the target node and the reference nodes will be found by exchanging the IP packets with time stamps of Time of Departure (ToD) and Time of Arrival (ToA). We have to keep record of the values of ToA and ToD of each transmission of a packet to find out the transmission time which is the difference between ToA and ToD. For each transmission we can use a set of packets with a known number of packets. We can consider N number of packets such as $P_1, P_2,......, P_n$ will be send from one reference node to target node. For each packet we have to record the time of departure (TOD) such as $t_1 d_0, t_2 d_0,......, t_n d_0$ in the sender side. Again, on the receiving end, we have to record the time of arrival (TOA) such as $t_1 a_0, t_2 a_0,......, t_n a_0$. According to the TOD and TOA of each packet we have to calculate the time difference between TOD and TOA for each packet. Then, we have to calculate the average transmission time by equation 3.5.

$$Average\ Transmission\ Time = \sum_{i=1}^{n} \frac{(TOA_i - TOD_i)}{n} \quad \ldots\ldots (3.5)$$

After finding the average transmission time we have to calculate average propagation time by the following equation.

Avg. Propagation Time= Avg. Transmission Time- Avg. Processing Time …… (3.6)

We have to use a reference value as a propagation time that relates to an indexed range or distance. By this way, we can get the range d_0 between the reference node R_0 and the target node T. We need to repeat the above process two times for getting the ranges d_1 and d_2 from other two reference nodes R_1 and R_2 to the target node T respectively. After getting all the ranges between each reference nodes to the target node, we have to find out the average value of the ranges. Finally, by using the average value of the rages, we will find out the coordinate of the target location T(X, Y). As our algorithm will able to track the location of a target node continuously, we have to repeat the overall procedure as well as the computation continuously for predicting the location of the target node in the network. We have to consider each location computed is with respect to a relative origin which is fixed and it has a GPS location. We can even change the relative location to the origin to a GPS value.

3.3 Dynamic References

In this research, dynamic references are used for getting more accuracy through geometric triangulation method compare to the basic triangulation method. One of the major concerns of this research is to keep track the moving target as well as stationary target. As the target can move any direction in mobile Adhoc network, it requires dynamic references for applying proper geometry in triangulation method instead of stationary references. When the zone of the target will be predicted with respect a relative origin, reference nodes need to be selected based on their PL&T information. The nodes that have accurate PL&T will be perfect as a reference node to form geometry. The accuracy of the triangulation method mainly depends on the proper geometry of the zone prediction and on the reference node selection.

4. SIMULATION AND RESULT

The ZF algorithm is based a predictive polynomial of nth order where n has to be at least 3. It has been implemented is simulation using MATLAB. The general polynomial for the vector distance of the object from a reference point R1 that is shown is Figure 1 is defined by equation 4.1. Reference point is basically a starting point since the polynomial is self-embedded.

$$X(t) = \sum_{k=0}^{n} a_k t^k \ldots\ldots\ldots\ldots (4.1)$$

Where n=order of the polynomial, a_k is the coefficient of the kth order of t, and t is the time at which object is located. X(t) describes the trajectory of the moving object in time.

$$d_j = \frac{\left| \Pi_{\substack{i=o \\ i \neq j}}^{m-1} (t_m - t_i) \right|}{\left| \Pi_{\substack{i=0 \\ i \neq j}}^{m} (t_j - t_i) \right|} \ldots\ldots\ldots\ldots (4.2)$$

Where d_j are the coefficients created for use in computing the vector distance X(m) of the object from the reference point R1 in order to continuously track the object based on the m immediately preceding locations.

$$X(t_m) = \sum_{k=0}^{m-1} d_k * X(t_k) \ldots\ldots\ldots (4.3)$$

To demonstrate the accuracy of the algorithm, we make the following assumption in equation (4.1): n=3; a_3=1; a_2=-1; a_1=2 and a_0=5. Then,

$$X(t) = t^3 - t^2 - 2t + 5 \ldots\ldots\ldots(4.4)$$

$$d_j = \frac{\left| \prod_{\substack{i=0 \\ i \neq j}}^{3}(t_4 - t_i) \right|}{\left| \prod_{\substack{i=0 \\ i \neq j}}^{3}(t_j - t_i) \right|} \ldots\ldots\ldots\ldots(4.5)$$

$$X(t_4) = \sum_{k=0}^{3} d_k * X(t_k) \ldots\ldots\ldots(4.6)$$

4.1. Simulation Scenario

We have simulated two particular scenarios for tracking the position location of a target node. In first scenario, we assume that that target node moves along straight line that means target node does not change the direction of moving. In second scenario, we consider that the target node changes the direction of moving such that target node goes along through the spiral, trapezoidal, or any other way except straight line. For both cases, we have simulated the zone finding (ZF) algorithm and triangulation method to track the position location of a target node for its seven location point. For both cases, in triangulation method four mobile nodes are used in the simulation, where three of them act as reference nodes and one act as a target node. For each instance of time, we have found the zone of the location of the target and then we have placed three reference nodes around the zone by satisfying the condition (Discussed in chapter 3) for forming three triangles where each triangle consists of two reference node and the target node. . After forming three triangles, we have sent a particular number of IP packets from each reference node to the target node. By keeping the records value of TOA and TOD of each transmission, we have find out the average transmission time for a particular number of packets from each reference node to the target node by applying equation 3.5. From the average transmission time, we have find out the propagation time by applying equation 3.6.

4.2 Result for Scenario One

According to the scenario one, to track the position location of a target, we have run the simulation for zone finding algorithm and for triangulation algorithm. The simulation process and the result we have found are discussed below.

ZF Algorithm: To execute simulation for zone finding, we used MATLAB as a simulation tool and used equations 4.1 to 4.6 for the above example. Equation 4.1 and 4.2 are used for continuously compute the fifth location (for the case n=3) of the target based on previous four location. As the target moves along the trajectory, when n goes higher, it will compute the next location by using the higher number of previous location. Table 1 illustrates the accuracy of the algorithm for the case n=3; a_3=1; a_2=-1; a_1=2 and a_0=5 based on our assumption. The first column of Table 1 shows the instance of time t that means on that particular time where the target is located. Second column presents the theoretically value of the distances of the target for each instance of time t and third column shows the predicted location distance that we have got from simulation by implementing the iterative process based on having four previous location. We have also find out the error between theoretical value and predicted value of trajectory location of the target that is presented in column four of Table 1.

Table 1. Illustration of the above example data.

Time	Theoretical trajectory location	Predicted Location distance using Iterative Process	% Error
0	5	-	-
1	7	-	-
2	13	-	-
3	29	-	-
4	61	61	0
5	115	115	0
6	197	197	0
7	313	313	0
8	469	469	0
9	671	671	0
10	925	925	0

The simulation of the algorithm shows high accuracy of prediction of the target location as shown in Table 1. We have plotted the true path (contains the actual locations of the target) and the ZF predicted path of the first 10 locations (over time) of the target using the polynomial $X(t) = t^3 - t^2 + 2t + 5$ (above).

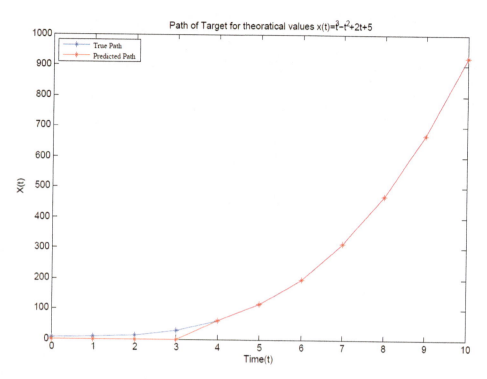

Figure 4. Illustration of accuracy of ZF algorithm

Figure 4 shows the true locations of the target node and compares it with the locations of the predicted value. It shows that the predicted locations at different time instances match closely with those of the true locations. The locations at each time instance is derived from the initial reference point and projects how far the target has moved at each time instance in meters.

Finding the Co-ordinates of the predicted locations: We need to convert the Distance between the initial location point and the current location point of the target node at each instance of time into [X, Y] coordinates. We used equation 3.2 and equation 3.4 (implemented in MATLAB) to convert the distances to the 2-D coordinates. We assume that the first location point's coordinates is L_1 (0, 0) which makes 30 degree angles along with the X-axis. Table 2 shows the simulation results of findings the X-Y coordinates of the first 7 predicted location points below.

Table 2. X-Y coordinates of first 7 predicted location point

Location Point (Number)	Predicted Location (m)	X-Coordinate	Y-Coordinate
1	61	25.1147	14.5000
2	115	52.8275	30.5000
3	197	99.5929	57.5000
4	313	170.6070	98.5000
5	469	271.0659	156.5000
6	671	406.1659	234.5000
7	925	581.1030	335.5000

Triangulation for location prediction within each zone: Once the locations within the zones are identified, the triangulation process is executed to find the exact location. We have used NS-2 as a simulation tool for performing the simulation of triangulation process that consists of several steps. In this simulation, we have found the locations of a target for 7 instance of time for a particular example that we have discussed in section 4.1. We run the simulation for finding 7 location point of a target node. Every time when a target node changes its location, it also dynamically changes all reference nodes' position by satisfying all the condition for making good geometry on triangulation. By applying equation 3.5, we have calculate average Transmission Time, Average Propagation Time for each reference node such as R1, R2, R3 to the target node for 7th location.

We have used a reference value as a propagation time to an indexed range. Here we assume,

1 ns=1 foot or 1ns=.3048 meter

From this assumption, we have determined the ranges from each reference node to the target location. After getting all three ranges for a target location, we have computed the average value of three ranges for a location point. Table 3 shows the average value of the ranges for 7 location point.

Table 3. Avg. range of the target node from three reference node for 7 location point

Target Location Point	Average Range (m)
1	9.3113379
2	20.6330321
3	43.5823384
4	82.7503576
5	141.806069
6	225.544281
7	335.408222

Then we have determined the coordinates of the target location by using the average value of ranges. Finally, by taking the average value of the computed coordinates, we have selected the final value of the target coordinates for 7 target location point. Table 4 shows the final value of X-Y coordinates of the target for 7 location point.

Table 4. Value of X-Y coordinates of the target node for 7 location point.

Target Node location Point	X-Coordinate	Y-Coordinate
1	25.1147	14.5
2	52.9018333	30.5363333
3	99.7072667	57.3482
4	170.710933	98.1731
5	271.156767	156.008033
6	406.166967	233.894367
7	581.151233	334.7497

According to the value of X-Y coordinate we have plotted a graph in Figure 5 that shows the 7 location points of the target node those are tracked through triangulation method and zone finding method.

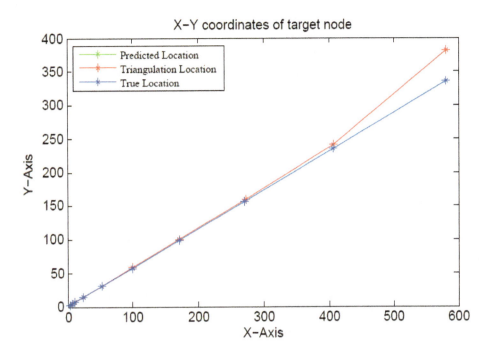

Figure 5. Location Points of the target node those are tracked through triangulation method and zone finding method.

4.3 Results for Scenario Two

According to the scenario one, to track the position location of a target, we have run the simulation for zone finding algorithm and for triangulation algorithm. The simulation process and the result we have found are discussed below.

ZF Algorithm: Here, we have considered that the target node moves along a spiral path. Since our ZF algorithm is based on a polynomial equation, it will predict the position location of the target node along with the straight line. Therefore, when the target node changes its direction, through only this ZF algorithm we cannot predict the position location correctly. For this reason, we need to apply both the ZF algorithm and the triangulation method for tracking the location point accurately. We have followed the same procedure to simulate the ZF algorithm for this scenario.

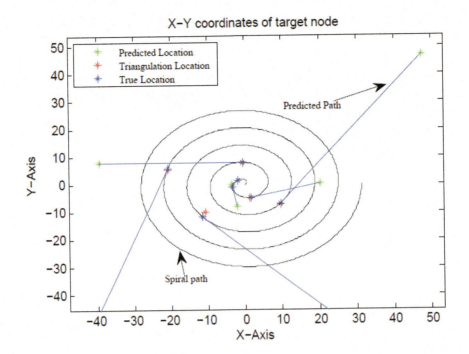

Figure 6. Location points of the target node those are tracked through triangulation method and zone finding method.

Triangulation for location prediction within each zone: Following the same procedure that we have already discussed in 4.4.a. section, we have simulated the triangulation method for find the position location of the target node for its seven location point along the spiral path. Figure 6 shows the location points of the target node those we have found through our implemented ZF algorithm and Triangulation method.

4.4. Performance Analysis:

Since we have the true value of X-Y coordinates of a target node's location, we have found out the percentage of error of the triangulation value through equation 4.7. Figure 7 shows the graph of the percentage of the error of the distances according to the true distance value for 7 location point of the target for scenario one. X-axis presents the percentage of the error and Y-axis presents the average distances between the reference nodes and the target node.

$$Error = \left(\frac{|(True\ Value - Triangulation\ Value)|}{True\ Value}\right) * 100 \quad \ldots\ldots\ldots (4.7)$$

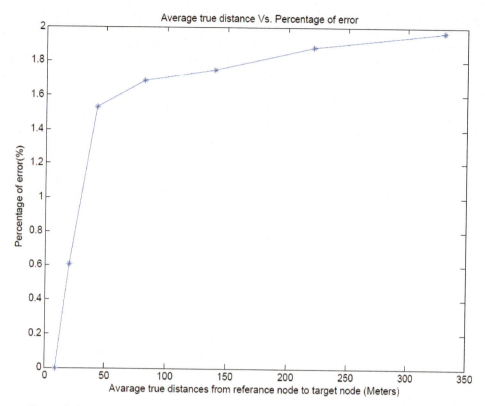

Figure 7. Percentage of the error of the measured distances Vs True distances of the target

We can see that, percentage of error is increasing gradually with average distance value from reference nodes to target node.

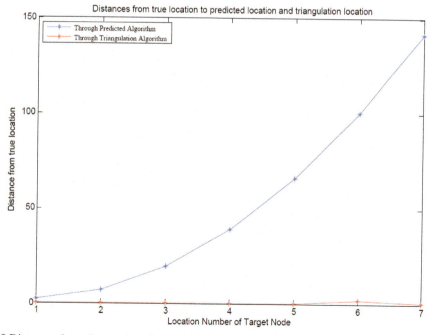

Figure 8 Distances from the true locations of the target node for 7 location points to its predicted location point and triangulation location point.

Figure 8 shows the graph where we can see the distances from true location point of a target node to its predicted location point and triangulation location point for second scenario. We can see that, in spiral way, the locations those we have found through the ZF algorithm are far away from the true locations of a target node. Therefore, to get the location point accurately, we need to apply triangulation method that help to predict the zone by using reference nodes when target node changes its direction of moving.

5. CONCLUSION

In this research, the design of a real time dynamic Position Location & Tracking (PL&T) system which is based on integrated predictive zone finding and triangulation method and its performance is shown. The zone finding of a target node is achieved using a predictive algorithm based on knowing previous good PL&T locations. Once the zone is predicted, the triangulation uses three references with good PL&T data to be placed around the target node such that the geometry is maintained properly for highly accurate prediction of the PL&T location of the target node. We use three references for two dimensional PL&T location. The same thing can be applied for deriving a three dimensional PL&T location of a target node using four references. This paper limits the discussions only to two dimensional PL&T location computation. The performance results are provided in this paper which shows highly accurate prediction of the PL&T location for mobile target nodes.

6. ACKNOWLEDGEMENT

This project is partially supported by funding from the National Science Foundation under the research grant NSF 0931679 and support of Research Assistantship from the Department of Electrical and Computer Engineering, Prairie View A&M University. The authors acknowledge the Mr. Golam Khan in the simulation assistance.

REFERENCES:

[1] N. Shakhakarmi & D. R. Vaman, (2012) "Real Time Position Location & Tracking (PL&T) using Prediction Filter and Integrated Zone Finding in OFDM Channel", *WSEAS transactions in Communications*, ISSN:1190-2742, Issue 7, Volume 11.

[2] L. Xiaofeng, W. Fletcher, L. Ian, L. Pietro, & X. Zhang, (2008) "Location Prediction Algorithm for Directional Communication", Wireless Communications and Mobile Computing Conference, IWCMC '08. International, ISBN: 978-1-4244-2201-2, IEEE Publisher .

[3] N. Shakhakarmi & D. R. Vaman, (2012) "Dynamic PL&T using Two Reference Nodes Equipped with Steered Directional Antenna for Significant PL&T Accuracy", Wireless Telecommunications Symposium 2012, London, UK.

[4] N. Shakhakarmi & D. R. Vaman, (2010) "Distributed Position Localization and Tracking (DPLT) of Malicious Nodes in Cluster Based Mobile Ad hoc Networks (MANET)", WSEAS transactions in Communications, ISSN: 1109-2742, Issue 11, Vol. 9.

[5] L. Xiaofeng, W. Fletcher, L. Ian, L. Pietro & X. Zhang, (2008) "Location Prediction Algorithm for Directional Communication", Computer Laboratory, University of Cambridge, U.K, and College of Computer Science, Beijing University of Aeronautics and Astronautics, China, IWCMC.

[6] R. Michael Buehrer, Tao Jia & Benton Thompson, (2010) "Cooperative Indoor Position Location using the Parallel Projection Method", 2010 international conference on indoor positioning and indoor navigation (ipin), zurich, switzerland.

[7] N. Malhotra, M. Krasniewski, C. Yang, S. Bagchi & W. Chappell, (2005) "Location Estimation in Ad-Hoc Networks with Directional Antennas", School of Electrical & Compute Engineering, Purdue University, West Lafayette, ICSCS.

[8] R. Siuli, C. Sanjay, B. Somprakash, U. Tetsuro, I. Hisato & O. Sadao, (2005) "Neighborhood Tracking and Location Estimation of Nodes in Ad hoc Networks Using Directional Antenna: A

[9] B. Zhang & F. Yu, (2010) "Low-complex energy-efficient localization algorithm for wireless sensor networks using directional antenna," Department of Integrated Electronics, Shenzhen Institutes of Advanced Technology, IET Commun., Vol. 4, pp. 1617–1623.

[10] P. Veeranath, Dr.D.N.Rao, Dr.S.Vathsal & N.Bhasker, (2013) "Reducing Multipath Effects in Indoor Channel for Analysis of GPS/Pseudolite Signal Acquisition", International Journal of Scientific and Research Publications, Vol. 3, ISSN 2250-3153, url:" www.ijsrp.org".

[11] K. Yedukondalu1, A. D. Sarma & V. Satya Srinivas, (2011) "Estimation and mitigation of gps multipath interference using adaptive filtering", Progress In Electromagnetics Research M, Vol. 21, pp.133–148.

[12] R. Ramanathan, (2005) "Antenna Beamforming and Power Control for Ad Hoc Networks", Mobile Adhoc Networking , BBN Technologies, Cambridge, Massachusetts, DOI: 10.1002/0471656895.ch5, url:" http://www.ir.bbn.com/~ramanath/pdf/wiley-bookchap.pdf".

[13] R. Ramanathan, J. Redi, C. Santivanez, D. Wiggins & S. Polit, (2005) "Ad Hoc Networking with Directional Antennas: A Complete System Solution", IEEE J. Sel. Areas Commun., vol. 23, no.3, pp. 496-506.

[14] Gilbert, J.M. & Won-Joon Choi, (2005) "MIMO Technology for Advanced Wireless Local Area Networks", Design Automation Conference, ISBN:,1-59593-058-2, pp 413 – 415.

[15] A. Pauraj, R. Nabar & D. Gore, (2003) "Introduction to Space-Time Wireless Communications", Cambridge University Press, Cambridge.

[16] Omnidirectional vs Directional, url "http://www.cisco.com/c/en/us/support/docs/wireless-mobility/wireless-lan-wlan/82068-omni-vs-direct.html".

Authors

Sharmistha Khan got the B.Sc. degree in Computer Science from American International University-Bangladesh (AIUB), Dhaka, Bangladesh in 2006, and the M.S. degree in Electrical Engineering from Tuskegee University, Tuskegee, AL in 2011, respectively. She is currently pursuing her PhD degree in the department of Electrical and Computer Engineering at the Prairie View A & M University, Prairie View, TX. She is working under the supervision of Prof. Dhadesugoor R. Vaman at Prairie View A & M University. Her research area includes Mobile Adhoc Network, Cognitive Radio Networks, Sensor Networks, Mobile WiMAX Technology, Handoff Management, and Handoff Performance and Decision Making Algorithms for Broadband Wireless Networks. She is a member of the CEBCOM group at the Prairie View A & M University.

Dhadesugoor R. Vaman is Texas Instrument Endowed Chair Professor and Founding Director of ARO Center for Battlefield Communications (CeBCom) Research, ECE Department, Prairie View A&M University (PVAMU). He has more than 38 years of research experience in telecommunications and networking area. Currently, he has been working on the control based mobile ad hoc and sensor networks with emphasis on achieving bandwidth efficiency using KV transform coding; integrated power control, scheduling and routing in cluster based network architecture; QoS assurance for multi-service applications; and efficient network management. Prior to joining PVAMU, Dr. Vaman was the CEO of Megaxess (now restructured as MXC) which developed a business ISP product to offer differentiated QoS assured multiservices with dynamic bandwidth management and successfully deployed in several ISPs. Prior to being a CEO, Dr. Vaman was a Professor of EECS and founding Director of Advanced Telecommunications Institute, Stevens Institute of Technology (1984-1998); Member, Technology Staff in COMSAT (Currently Lockheed Martin) Laboratories (1981-84) and Network Analysis Corporation (CONTEL)(1979-81); Research Associate in Communications Laboratory, The City College of New York (1974-79); and Systems Engineer in Space Applications Center (Indian Space Research Organization) (1971-1974). He was also the Chairman of IEEE 802.9 ISLAN Standards Committee and made numerous technical contributions and produced 4 standards. Dr. Vaman has published over 200 papers in journals and conferences; widely lectured nationally and internationally; has been a key note speaker in many IEEE and other conferences, and industry forums. He has received numerous awards and patents, and many of his innovations have been successfully transferred to industry for developing commercial products.

13

IMPACT OF RANDOM MOBILITY MODELS ON OLSR

P. S. Vinayagam

Assistant Professor, Department of Computer Science,
Pondicherry University Community College, Puducherry, India

ABSTRACT

In ad hoc networks, routing plays a pertinent role. Deploying the appropriate routing protocol is very important in order to achieve best routing performance and reliability. Equally important is the mobility model that is used in the routing protocol. Various mobility models are available and each can have different impact on the performance of the routing protocol. In this paper, we focus on this issue by examining how the routing protocol, Optimized Link State Routing protocol, behaves as the mobility model is varied. For this, three random mobility models, viz., random waypoint, random walk and random direction are considered. The performance metrics used for assessment of Optimized Link State Routing protocol are throughput, end-to-end delay and packet delivery ratio.

KEYWORDS

OLSR, Mobility model, Random Waypoint, Random Walk, Random Direction

1. INTRODUCTION

Wireless networks can be classified into infrastructure based and infrastructure less networks. In the case of infrastructure based networks, Access Points are used for communication. They act as routers for the nodes within their communication range. Whereas, in infrastructure less networks, also known as, ad hoc networks, nodes act as routers. That is, such networks do not have predesignated routers and nodes connect in a dynamic manner. A node cannot connect to all other available nodes using single hop as the transmission range of nodes is limited and hence data is transmitted using multi hop. A mobile ad hoc network (MANET) is a type of ad hoc network in which nodes can change locations. It is a self configuring infrastructure less network of mobile devices connected by wireless links [1].

The routing protocols in MANET are broadly classified into three categories, namely, proactive protocols, reactive protocols and hybrid protocols. Proactive protocols, also known as table-driven protocols, maintain routing information in the routing table of each node. The routing table is populated in a proactive manner and the routing table information is transmitted to other neighboring nodes at fixed time intervals. Few examples of proactive routing protocols are Optimized Link State Routing (OLSR) Protocol, Destination-Sequenced Distance-Vector (DSDV) Routing Protocol.

Reactive routing protocols are also known as demand driven protocols. In these protocols, prior route information to other nodes is not maintained. Whenever a node (source node) needs to transmit data to a destination node, the route is determined on demand. The node initiates a route discovery process only if it has data destined to a particular node. For other nodes for which no data is to be transmitted, routes are not computed. Examples of reactive routing protocols are

Dynamic Source Routing (DSR) Protocol, Ad hoc On-Demand Distance Vector (AODV) Routing Protocol etc.

The third category, hybrid routing protocols, are a combination of both proactive and reactive routing protocols. For example, proactive routing may be used to communicate with neighbors and reactive may be used to communicate with distant nodes. Examples of hybrid routing protocols are Zone Routing Protocol (ZRP), Core Extraction Distributed Ad Hoc Routing (CEDAR) Protocol etc.

The performance of the varied routing protocols may be dependent on various factors; one such key parameter that could impact routing protocols is the mobility model. The mobility model is the one that is used to describe the pattern in which mobile users move. It also describes how the location and velocity of the nodes change over time. Based on the mobility model being used, the performance of a routing protocol can vary.

In this paper, we assess the impact of the various random mobility models on OLSR protocol using the performance metrics, throughput, end-to-end delay and packet delivery ratio. The rest of the paper is organized as follows. The OLSR protocol and the various mobility models are briefly discussed in section 2. Section 3 consolidates the related work on the performance of routing protocols using various mobility models. The simulation environment, performance metrics and the simulation results are discussed in section 4. Section 5 concludes the paper.

2. OLSR AND MOBILITY MODELS

2.1. Optimized Link State Routing (OLSR) Protocol

OLSR protocol is a proactive type of routing protocol. It uses Multipoint Relay (MPR) sets for routing. For each node, a set of its neighbor nodes that have symmetric links are selected as MPRs, which alone forward the control traffic. When a node is selected as multipoint relay, it announces this information in the control messages at periodic intervals. Using this, routes are formed from a given node to various destinations. Nodes that belong to MPR set cover all symmetric strict 2-hop neighbor nodes.

In OLSR, HELLO messages and topology control messages are used. HELLO messages are transmitted at regular intervals and they are never forwarded. The HELLO messages help in link sensing, neighbor detection and MPR selection signaling. Link-state information of each and every node is transmitted to all other nodes in the network via the topology control messages. This helps the nodes to compute their routing table. Topology control messages are sent using the MPRs.

2.2. Mobility Models

Mobility models are generally classified into five categories [2]. They are random mobility models, mobility models with temporal dependency, mobility models with spatial dependency, mobility models with geographic restrictions and hybrid mobility models. This classification is summarized in figure 1 [1, 2].

In random mobility models, the nodes move independently by choosing a random direction and speed. In the case of mobility models with temporal dependency, the movement of nodes is affected by their movement history. In the mobility models with spatial dependency, the movement of nodes is correlated in nature. If the mobility model limits the movement of nodes owing to streets or obstacles, then such models fall under mobility models with geographic restriction [1]. In hybrid mobility models, mobility models with spatial dependencies, temporal dependencies and geographic restrictions are integrated [2].

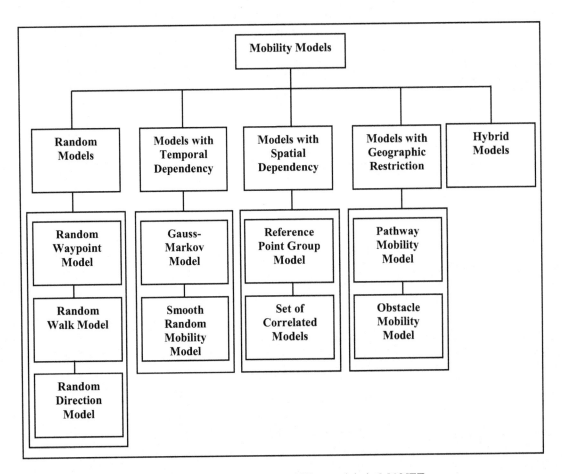

Figure 1. Classification of mobility models in MANET

Of the various mobility models, three random mobility models, viz., Random Waypoint, Random Walk and Random Direction model are considered in this study to assess the performance of OLSR under varying mobility pattern. In the next section we elaborate on these three mobility models.

2.2.1. Random Waypoint Mobility Model

In the Random Waypoint Mobility Model, a node selects a random position (x, y) in the simulation area. This point serves as the destination point. A velocity (v) is chosen from a uniformly distributed range [minspeed, maxspeed]. The node travels to the destination point with speed v. Upon reaching the destination point, the node pauses for a specified pause time. Then again the node repeats the above process by choosing a new destination and speed [3].

2.2.2. Random Walk Mobility Model

In random walk mobility model, nodes move by randomly choosing a speed and direction in constant time intervals (Δt). The speed is determined from the range [minspeed, maxspeed] and the direction $\theta(t)$ is chosen from the range [0, 2Π]. The node moves with the velocity vector (v(t) cosθ(t), v(t) sinθ(t)). When the node reaches the simulation boundary, it bounces back to the simulation area. The angle of bouncing is θ(t) or Π − θ(t). This effect is called as border effect. This model is also referred to as the Brownian Motion Mobility Model or Brownian Walk. Random Walk model can also be considered as Random Waypoint model with zero pause time [1, 3].

2.2.3. Random Direction Mobility Model

In the case of Random Direction Mobility Model, a node chooses a random direction uniformly within the range [0, 2Π]. The velocity is also chosen uniformly from within the range [minspeed, maxspeed]. Node then moves in the chosen direction until it arrives at the boundary of the simulation area. At this point the node pauses for a specified pause time and again selects a new direction from within the range [0, Π]. Since the node is on the boundary of the simulation area, the direction is limited to Π [3].

3. RELATED WORK

Various studies have been conducted so far to assess the performance of routing protocols in the context of different mobility models.

A survey of various mobility models in cellular networks and multi-hop networks has been carried out in [4]. Reference point group mobility model has been applied to two different network protocol scenarios, clustering and routing. The performance of the network with different mobility patterns and for different protocols has been studied. It is found that the performance of the protocols varies with different mobility patterns used. In AODV and Hierarchical State Routing (HSR), the throughput improves when the communications are restricted within the scope of a group, whereas DSDV is not affected by group mobility and localized communications.

The performance of DSR and AODV with reference to varying network load, mobility and network size has been analyzed in [5]. DSR is found to outperform AODV in small networks with low load and/or mobility, whereas AODV is more efficient with increased load and/or mobility. However routing load is less in DSR compared to AODV. It is observed that using congestion-related metrics and aged packets removal can improve the performance of both DSR and AODV. The authors conclude that the interplay between routing and MAC layers affects the performance of the protocols significantly.

The authors in [6] have studied the performance of DSR using random walk, random waypoint, random direction and reference point group mobility models. It is observed that the performance of the protocols varies with the different mobility models and even with the same mobility model but with different parameters.

To create realistic movement scenarios, the authors in [7] have used obstacles to restrict node movement and wireless transmissions. Also pathways have been constructed using Voronoi path computation. It is observed that the obstacles and pathways affect the performance of the protocols. AODV has been used to study the performance of routing protocol using obstacle model and random waypoint mobility model. It is observed that the mobility model chosen affects connectivity of the nodes, network density, packet delivery and routing overhead.

Various protocol independent metrics have been proposed to capture mobility characteristics including spatial and temporal dependence and geographic restrictions in [8]. Random waypoint, group mobility, freeway and Manhattan mobility models have been used to study the performance of DSR, AODV and DSDV. The results indicate that different mobility patterns affect the performance of the routing protocols. The mobility pattern influences the connectivity graph which affects the performance of the protocol. A preliminary investigation of the common building blocks of routing protocols was also attempted.

In [9] the authors introduced GEMM, a tool to generate mobility models that happen to be more realistic and heterogeneous. They have compared the performance of AODV, OLSR and ZRP with various mobility scenarios. It is observed that GEMM can generate more realistic mobility patterns than random waypoint. Mobility pattern influences the performance of the protocols.

In [10] the authors observe that high node degree can be both an asset and a liability. On one hand high node degree can affect scalability. On the other hand high node degree provides multiple routing options.

The performance of DSR and AODV using various mobility models has been studied in [11]. They observe that the mobility pattern affects the performance of routing protocols and that mobility metrics, connectivity and performance are related. When relative speed increases with similar average spatial dependency, there is decrease in link duration and hence routing overhead increases and throughput decreases. In the case of similar average relative speed, the spatial dependence increases and the link duration increases, and hence there is an increase in the throughput and a decrease in the routing overhead. DSR and AODV have highest throughput and least overhead when reference point group mobility model is used. They conclude that mobility pattern influences the connectivity graph which impacts the performance of the routing protocol.

In [12] the performance of AODV routing protocol using pursue group and random based entity mobility models is studied. Pursue group mobility model has performed better than random based entity model.

The effect of mobility models on the performance of the protocols has been analyzed in [13] both analytically and through simulation. They present an analytical framework for the characterization of link and use it to describe lifetime of the path and stability of the topology. The framework describes link, path and topology dynamics as a function of node mobility. They find that there is a diminishing effect on the protocols with increase in mobility.

The performance of On-Demand Multicast Routing Protocol, Multicast Ad hoc On-Demand Distance Vector Routing Protocol and Adaptive Demand driven Multicast Routing Protocol have been studied using random way point, reference point group and Manhattan mobility models in [14]. It is evident from their results that with different mobility patterns the ranking of protocols differ.

In [15] the authors have used Levy-Walk mobility model and Gauss-Markov model to compare Adhoc On Demand Multipath Distance Vector (AOMDV) and OLSR routing protocols. They observe that AOMDV gives higher packet delivery and throughput, whereas OLSR has less delay and routing overhead in the context of varying node density. Also OLSR performs better than AOMDV under Levy-Walk mobility model.

The authors in [16] have studied the performance of AODV, DSR, DSDV, OLSR and Dynamic MANET On-Demand (DYMO) routing protocols using various mobility models. A fair comparison of the capabilities and limitations of different mobility patterns has been attempted.

The performance of AODV and DSR using reference region group mobility model has been examined in [17]. The reference region group mobility model is used to mimic group operations such as group partitions and mergers. It is found that the group partitions have an impact on the performance of the routing protocols. Frequent group partitions can downgrade the performance of both the routing protocols under consideration. Comparatively AODV is able to tackle better the group operations than DSR. Further AODV is more adaptive to high speed environment, while DSR is more suitable for networks with less mobility.

The authors in [18] have compared different hierarchical (position and non-position based) protocols using different mobility models. Position based routing protocols have performed better compared to their counterparts with reference to packet delivery. It is observed that non-position based routing protocols provide low packet delivery ratio and high packet loss. Further the authors conclude that the network performance can be enhanced in the presence of a recovery mechanism.

The authors in [19] have studied the impact of swarming behavior of nodes on the performance of routing protocols both analytically and through simulation, and they have also proposed a Markov swarm mobility model to characterize time-dependent changes in the network topology. They observe that owing to swarm movement of nodes in a collaborative manner, the routing overhead and average end-to-end delay is significantly reduced.

In [20] the authors have evaluated structured and unstructured content discovery protocols with various mobility models. It is evident from their work that movement patterns which exhibit more uniform distribution of nodes provide better efficiency. Limitations reduce the efficiency of the network. Increase in node speed does not have a considerable effect on path availability. They conclude that path availability is the most important factor affecting the efficiency of content delivery protocols. Hence mobility is not of much concern in implementation of efficient overlay networks. In the case of structured protocols which are not efficient for MANETs, the mobility has a negative effect on performance. Performance is dependent on stability and optimality of overlay in the case of structured protocols. In the case of unstructured protocols, alternative paths can be replaced in the case of link failure and hence unstructured protocols perform better.

Three distinctive mobility models in terms of node movement behaviour have been studied by the authors in [21]. A new measurement technique called probability or route connectivity has been used. This metric quantifies the success rate of route established by the routing protocol.

The performance of DSR, Location Aided Routing (LAR) and Wireless Routing Protocol (WRP) have been studied with reference to random waypoint mobility model, reference point group mobility model, Manhattan Grid mobility model and Gauss-Markov mobility model in [22]. It is found that the performance of routing protocols varies significantly with the mobility model being used and also the node speed affects the network performance. Location-based routing protocols exhibit good performance with various mobility patterns.

The authors in [23] have compared the performance of AODV, OLSR and DSDV with respect to reference point group mobility and random waypoint mobility models. It is found that, in the case of random waypoint mobility model, AODV shows maximum packet delivery ratio, least routing load and MAC load. As mobility increases, OLSR performs better with respect to delay. In reference point group mobility model, AODV has higher packet delivery ratio and lowest routing load, whereas OLSR exhibits least delay and maximum MAC load.

The performance of AODV and DSDV using random waypoint, reference point group mobility, Freeway and Manhattan mobility models have been analyzed in [24]. It is observed that AODV has stable performance in all the mobility models studied. It performs best with group mobility model and freeway model. Performance of DSDV is unstable with random waypoint, Freeway and Manhattan mobility models. Performance is best in the case of Reference Point Group Mobility model for both the protocols. AODV has high throughput and low end-to-end delay, whereas both AODV and DSDV have relatively same packet delivery ratio. DSDV suffers from high packet drop compared to the other protocol under consideration.

In [25] the authors have compared AODV, DSR, OLSR, DSDV and Temporally Ordered Routing Algorithm (TORA) routing protocols using reference point group mobility (RPGM), column mobility model (CMM) and random waypoint (RWP) mobility models. The results show that reactive protocols perform better than proactive protocols with reference to packet delivery ratio, end-to-end delay, normalized routing load and throughput. OLSR has got the minimum delay whereas it is maximum in the case of DSR. Throughput is found to be maximum in AODV. DSDV performs better in the case of packet dropper whereas it is worst in the case of AODV. Increasing the number of nodes impacts the performance which varies based on protocols and mobility models. DSR shows degradation as the number of nodes increases. TORA's performance is very minimal.

In [26] the authors have studied AODV, DSR, LAR and OLSR routing protocols with random waypoint, reference point group mobility, Gauss Markov and Manhattan Grid mobility models. They report that there is significant impact on the performance of the routing protocols based on the mobility model being used. The protocols have exhibited considerable difference for different mobility models. The choice of the mobility model has most impact on DSR and OLSR.

The performance of AODV, OLSR and gathering-based routing protocol (GRP) using random waypoint and vector mobility models has been evaluated in [27]. OLSR performs better in terms of throughput and end-to-end delay. In all the routing protocols studied, vector mobility model outperforms random waypoint mobility model.

The effect of random waypoint mobility model and group mobility model for both constant bit rate and variable bit rate traffic has been studied in [28]. They have used AODV, OLSR and ZRP for comparison. With respect to throughput, end-to-end delay and jitter, OLSR performs better than AODV and ZRP. Performance of ZRP is found to be the least among the three protocols.

In [29] the authors have studied the performance of random waypoint and vector mobility model with reference to AODV, OLSR and GRP. They have concluded that OLSR performs better in terms of throughput and end-to-end delay. It is also observed that AODV has lesser network load in both the mobility models used for simulation.

The performance of OLSR, TORA and ZRP with reference to random waypoint mobility model, reference point group mobility model and Manhattan mobility model has been analyzed in [30]. They have concluded that different factors such as pause time, node density and scalability affect the performance and efficiency of the protocols. They also state that no single protocol gives optimum efficiency.

In [31] the authors have evaluated the performance of AODV, DSR and DSDV with respect to different network loads and various mobility models. They found that the performance of routing protocols varies with different mobility models. DSR protocol exhibits better performance with random waypoint mobility model but in the case of Manhattan Grid Mobility model its performance is fair. The end-to-end delay is lowest in the case of RPGM model and it exhibits best performance in DSDV. The authors observed that DSR routing protocol with random waypoint mobility model is better compared to the other combinations.

The authors in [32] have analyzed the performance of AOMDV using random waypoint, random direction and probabilistic random walk mobility models. Their results show that packet delivery ratio decreases with increasing node mobility in all the mobility models. Average end-to-end delay is also affected with varying node speed. With reference to packet delivery ratio and average end-to-end delay, AOMDV performs better with random waypoint mobility model.

In [33] the authors have examined the performance of AODV, DSR, OLSR, DSDV and TORA with reference to reference point group mobility model, column mobility model and random walk mobility model. It is found that delay is least in the case of OLSR and maximum in DSR. AODV shows high throughput whereas DSDV performs better with reference to packet dropper. Performance of DSR declines with increase in the number of nodes, while that of TORA is very poor compared to the other protocols under consideration.

The impact of mobility models and traffic patterns on AODV, DSDV and OLSR has been studied using both CBR and TCP traffic patterns with respect to reference point group and Manhattan Grid mobility models in [34]. The performance metrics used are packet delivery ratio, throughput and end-to-end delay. It is observed that the relative ranking of protocols varies based on the mobility model, node speed and the traffic patterns used. The authors conclude that AODV performed better with TCP-Vegas compared to the two other protocols under consideration. Also the performance was better with TCP traffic patterns compared to CBR traffic pattern. The end-to-end delay was better in DSDV and OLSR when CBR traffic pattern and reference point group mobility model is used.

Performance of AODV, OLSR and TORA using random walk mobility model and random waypoint mobility model is compared in [35]. Different types of traffic have been used to arrive at the results. They conclude that OLSR gives best performance in terms of throughput and load, but has higher delay than the other two protocols. In the case of mobility model, random waypoint mobility model is found to be better than random walk mobility model in all the three routing protocols that have been compared.

Random waypoint and reference point group mobility models have been used to study the performance of DSR, OLSR and TORA in [36]. The results show that reactive protocols are better than proactive protocols in terms of packet delivery fraction, end-to-end delay and throughput. DSR has performed better than OLSR and TORA, whereas performance of TORA is the least among the three protocols considered. OLSR has exhibited average performance in both the mobility models whereas DSR has performed better in random waypoint mobility model.

In this paper, we assess the impact of random waypoint, random walk and random direction mobility models on OLSR protocol with reference to performance metrics, viz., throughput, end-to-end delay and packet delivery ratio.

4. SIMULATION RESULTS

4.1. Simulation Environment and Performance Metrics

Simulations have been carried out using NS3, a discrete event network simulator [37]. Random waypoint, random walk and random direction mobility models have been used to evaluate their impact on OLSR. Simulation is run for a total of 300 seconds using 50 nodes spread over an area of 1000m x 1000m. The speed of the nodes is varied from 10m/s to 50m/s in steps of 10m/s and the pause time is 10 seconds. The packet size is 512 bytes and the channel capacity is 5.5 Mbps. The MAC protocol used is 802.11b.

Three performance metrics, viz., throughput, end-to-end delay and packet delivery ratio (PDR) are examined. Throughput refers to the average rate at which data packet is delivered successfully from one node to another. It is usually measured in bits per second. End-to-end delay is the time taken for a data packet to reach its destination. It is the difference between the time a packet is sent and the time the packet is received. Packet delivery ratio is the ratio of data packets successfully delivered to the destinations to those generated by the sources. It is calculated by dividing the number of packets received by the destination by the number of packets sent by the source.

4.2. Result Analysis

4.2.1. Performance of OLSR using the three mobility models over varying node speed

The simulation results obtained using OLSR with random waypoint, random walk and random direction mobility models over varying node speed are shown in figures 2, 3 and 4. Figure 2 presents the results of throughput for varying node speed from 10 m/s to 50 m/s. From the figure, it is evident that the performance of OLSR with respect to throughput using random waypoint and random walk mobility models is almost similar with very little difference. But as the node speed increases the throughput using random waypoint mobility model is found to be consistent whereas random walk shows decline in the throughput. In the case of random direction mobility model, as the node speed increases there is substantial drop in the throughput.

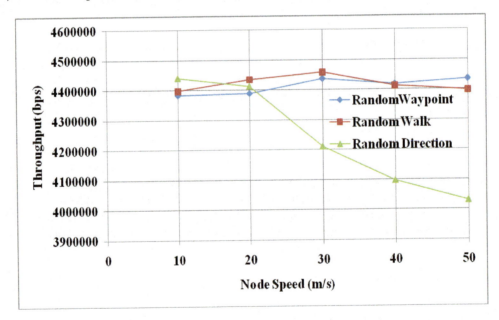

Figure 2. Throughput of OLSR using three mobility models over varying speed

The delay incurred by OLSR protocol using the three mobility models under consideration is shown in figure 3. With reference to end-to-end delay, the OLSR protocol using random waypoint mobility model exhibits least delay and it is consistent with increase in speed. In the case of random walk mobility model, delay is greater than random waypoint mobility model, but it is far better than random direction mobility model, which exhibits high end-to-end delay as the node speed increases.

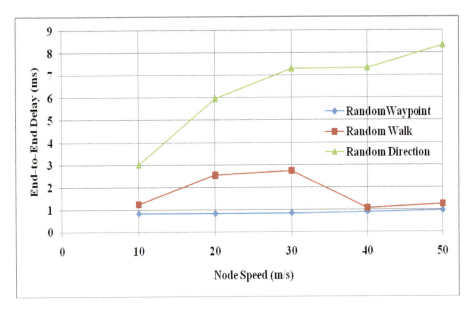

Figure 3. End-to-end delay in OLSR using three mobility models over varying speed

Figure 4 depicts the packet delivery ratio of OLSR protocol under the three mobility models. As is evident from the figure, random direction mobility model provides better packet delivery ratio than the other two mobility models, but at the cost of low throughput and high end-to-end delay. Among the other two mobility models, random walk provides better packet delivery ratio than random waypoint. The performance of random waypoint mobility model with reference to packet delivery ratio improves with increase in node speed, whereas random walk exhibits inconsistent packet delivery ratio with varying speed.

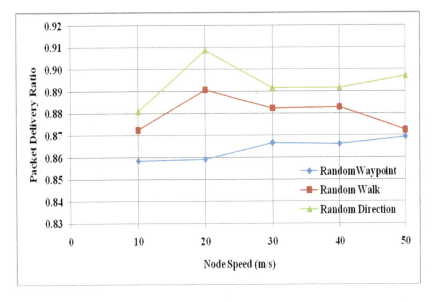

Figure 4. Packet Delivery Ratio in OLSR using three mobility models over varying speed

4.2.2. Performance of OLSR using the three mobility models over the simulation time

The results obtained for node speed 50m/s over the entire period of simulation time with respect to throughput, end-to-end delay and packet delivery ratio is depicted in figures 5, 6 and 7 respectively. As is evident from figure 5, random waypoint mobility model and random walk mobility model are comparatively similar with reference to throughput. But the throughput using random waypoint mobility model is consistent, while that of random walk shows gradual decline over time. End-to-end delay, as shown in figure 6, is lowest in the case of random waypoint mobility model and highest when random direction mobility model is used. Random walk is better than random direction with respect to end-to-end delay and with the passage of time decrease in end-to-end delay is observed. From the results of packet delivery ratio shown in figure 7, random direction seems to outperform the other two mobility models under consideration, but it exhibits low throughput and high end-to-end delay. Random walk mobility model provides better Packet Delivery Ratio than random waypoint mobility model.

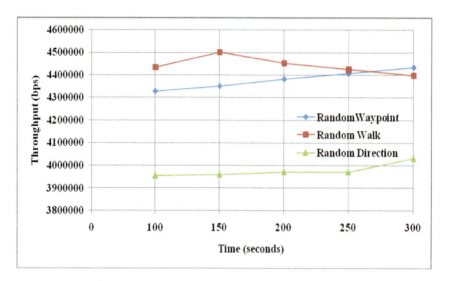

Figure 5. Throughput of OLSR using three mobility models over simulation time

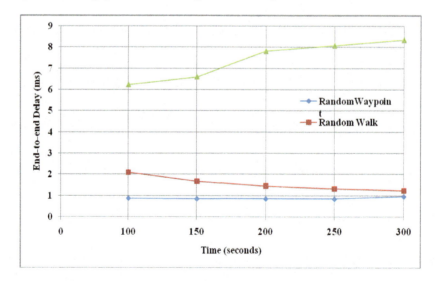

Figure 6. End-to-end Delay in OLSR using three mobility models over simulation time

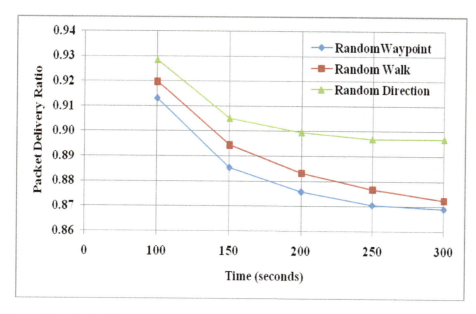

Figure 7. Packet Delivery Ratio in OLSR using three mobility models over simulation time

5. CONCLUSION

The impact of the various random mobility models, viz., random waypoint, random walk and random direction, on OLSR protocol with respect to throughput, end-to-end delay and packet delivery ratio has been examined. From the simulation results, it is clear that each of the mobility models outperforms the other two with respect to any one of the parameters throughput, end-to-end delay and packet delivery ratio. Considering the three parameters together, the performance of random direction mobility model does not seem to be better than the other two mobility models. It provides better packet delivery, but at the cost of lower throughput and higher end-to-end delay. As far as random walk and random waypoint is considered, OLSR with random waypoint provides good throughput and low end-to-end delay. But with respect to packet delivery ratio, random walk outperforms random waypoint mobility model.

It is evident from the results that the performance of OLSR under various metrics varies from one mobility model to another. There is significant impact of the mobility model on the routing protocol. In the future, random waypoint can be compared with group mobility models to see its effect on the routing protocol.

REFERENCES

[1] Fan Bai and Ahmed Helmy, "A Survey of Mobility Models," in *Wireless Adhoc Networks*. Available: http://www.cise.ufl.edu/~helmy/papers/Survey-Mobility-Chapter-1.pdf

[2] Nils Aschenbruck, Elmar Gerhards-Padilla, and Peter Martini, (2008) "A survey on mobility models for performance analysis in tactical mobile networks", *Journal of Telecommunications and Information Technology*, No. 2, pp. 54-61.

[3] Emre Atsan and Oznur Ozkasap, "A Classification and Performance Comparison of Mobility Models for Ad Hoc Networks," in *Ad-Hoc, Mobile, and Wireless Networks*, Springer-Verlag Berlin Heidelberg, 2006, pp. 444-457.

[4] Xiaoyan Hong, Mario Gerla, Guangyu Pei, and Ching-Chuan Chiang, (1999) "A Group Mobility Model for Ad HocWireless Networks", in *Proceedings of the 2nd ACM International Workshop on Modeling, Analysis and Simulation of Wireless and Mobile systems (MSWiM '99)*, pp. 53-60.

[5] Charles E. Perkins, Elizabeth M. Royer, Samir R. Das, and Mahesh K. Marina, (2001) "Performance Comparison of Two On-Demand Routing Protocols for Ad Hoc Networks", *IEEE Personal Communications*, Vol. 8, No. 1, pp. 16-28.

[6] Tracy Camp, Jeff Boleng, and Vanessa Davies, (2002) "A Survey of Mobility Models for Ad Hoc Network Research", *Wireless Communication & Mobile Computing*, Vol. 2, No. 5, pp. 483-502.

[7] Amit Jardosh, Elizabeth M. Belding-Royer, Kevin C. Almeroth, and Subhash Suri, (2003) "Towards Realistic Mobility Models For Mobile Ad hoc Networks", in *Proceedings of the 9th Annual International Conference on Mobile Computing and Networking (MobiCom '03)*, pp. 217-229.

[8] Fan Bai, Narayanan Sadagopan, and Ahmed Helmy, (2003) "The IMPORTANT framework for analyzing the Impact of Mobility on Performance Of RouTing protocols for Adhoc NeTworks", *Ad Hoc Networks*, Vol. 1, No. 4, pp. 383-403.

[9] Michael Feeley, Norman Hutchinson, and Suprio Ray, "Realistic Mobility for Mobile Ad Hoc Network Simulation", in *Ad-Hoc, Mobile, and Wireless Networks*, Springer-Verlag Berlin Heidelberg, 2004, pp. 324–329.

[10] Brent Ishibashi and Raouf Boutaba, (2005) "Topology and mobility considerations in mobile ad hoc networks", *Ad Hoc Networks*, Vol. 3, No. 6, pp. 762-776.

[11] Geetha Jayakumar and Gopinath Ganapathi, (2008) "Reference Point Group Mobility and Random Waypoint Models in Performance Evaluation of MANET Routing Protocols", *Journal of Computer Systems, Networks, and Communications*, Vol. 2008.

[12] S H Manjula, C N Abhilash, Shaila K, K R Venugopal, and L M Patnaik, (2008) "Performance of AODV Routing Protocol using Group and Entity Mobility Models in Wireless Sensor Networks", in *Proceedings of the International MultiConference of Engineers and Computer Scientists Vol II IMECS 2008*.

[13] Xianren Wu, Hamid R. Sadjadpour, J.J. Garcia-Luna-Aceves, and Hui Xu, (2008) "A hybrid view of mobility in MANETs: Analytical models and simulation study", *Computer Communications*, Vol. 31, No. 16, pp. 3810–3821.

[14] R. Manoharan and E. Ilavarasan, (2010) "Impact of Mobility on the Performance of Multicast Routing Protocols in MANET", *International Journal of Wireless & Mobile Networks (IJWMN)*, Vol. 2, No. 2, pp. 110-119.

[15] S. Gowrishankar, Subir Kumar Sarkar, and T.G. Basavaraju, (2010) "Analysis of AOMDV and OLSR Routing Protocols Under Levy-Walk Mobility Model and Gauss-Markov Mobility Model for Ad Hoc Networks", *International Journal on Computer Science and Engineering (IJCSE)*, Vol. 2, No. 4, pp. 979-986.

[16] F. Maan and N. Mazhar, (2011) "MANET routing protocols vs mobility models: A performance evaluation", in *Proceedings of the Third International Conference on Ubiquitous and Future Networks (ICUFN)*, pp. 179-184.

[17] Yan Zhang, Chor Ping Low, and Jim Mee Ng, (2011) "Performance Evaluation of Routing Protocols on the Reference Region Group Mobility Model for MANET", *Wireless Sensor Network*, Vol. 3, No. 3, pp. 92-105.

[18] Atta ur Rehman Khan, Shahzad Ali, Saad Mustafa, and Mazliza Othman, (2012) "Impact of mobility models on clustering based routing protocols in mobile WSNs", in *Proceedings of the 10th International Conference on Frontiers of Information Technology (FIT)*, pp. 366-370.

[19] Jun Li, Yifeng Zhou, Louise Lamont, F. Richard Yu, and Camille-Alain Rabbath, (2012) "Swarm mobility and its impact on performance of routing protocols in MANETs", *Computer Communications*, Vol. 35, No. 6, pp. 709-719.

[20] Mahmood Fathy, Kaamran Rahemifar, Hamideh Babaei, Morteza Romoozi, and Reza Berangy, (2012) "Impact of Mobility on Performance of P2P Content Discovery Protocols Over MANET", *Procedia Computer Science*, Vol. 10, pp. 642-649.

[21] Megat Zuhairi, Haseeb Zafar, and David Harle, (2012) "The Impact of Mobility Models on the Performance of Mobile Ad Hoc Network Routing Protocol", *IETE Technical Review*, Vol. 29, No. 5, pp. 414-420.

[22] Misbah Jadoon, Sajjad Madani, Khizar Hayat, and Stefan Mahlknecht, (2012) "Location and Non-Location based Ad-Hoc Routing Protocols using Various Mobility Models: A Comparative Study", *The International Arab Journal of Information Technology*, Vol. 9, No. 5, pp. 418-427.

[23] Neha Rani, Preeti Sharma, and Pankaj Sharma, (2012) "Performance Comparison of Various Routing Protocols in Different Mobility Models", *International Journal of Ad hoc, Sensor & Ubiquitous Computing (IJASUC)*, Vol. 3, No. 4, pp. 73-85.

[24] R. Mohan, C. Rajan, and N. Shanthi, (2012) "A Stable Mobility Model Evaluation Strategy for MANET Routing Protocols", *International Journal of Advanced Research in Computer Science and Software Engineering*, Vol. 2, No. 12, pp. 58-65.

[25] Sunil Kumar Kaushik, Kavita Chahal, Sukhbir Singh, and Sandeep Dhariwal, (2012) "Performance Evaluation of Mobile Ad Hoc Networks with Reactive and Proactive Routing Protocols and Mobility Models", *International Journal of Computer Applications*, Vol. 54, No. 17, pp. 28-35.

[26] K.P. Hrudya, Brajesh Kumar, and Prabhakar Gupta, (2013) "Impact of Mobility on Different Routing Approaches in MANETs", *International Journal of Computer Applications*, Vol. 67, No. 23, pp. 18-22.

[27] Narinder Pal and Renu Dhir, (2013) "Analyze the Impact of Mobility on Performance of Routing Protocols in MANET Using OPNET Modeller", *International Journal of Advanced Research in Computer Science and Software Engineering*, Vol. 3, No. 6, pp. 768-772.

[28] Shubhangi Mishra, Ashish Xavier Das, and A.K. Jaisawal, (2013) "Effect of Mobility and Different Data Traffic in Wireless Ad-hoc Network through QualNet", *International Journal of Engineering and Advanced Technology (IJEAT)*, Vol. 2, No. 5, pp. 364-368.

[29] Sohajdeep Singh and Parveen Kakkar, (2013) "Investigating the Impact of Random Waypoint AND Vector Mobility Models on AODV, OLSR and GRP Routing Protocols in MANET", *International Journal of Computer Applications*, Vol. 63, No. 3, pp. 30-34.

[30] Sunil Kumar Singh, Rajesh Duvvuru, and Amit Bhattcharjee, (2013) "Performance Evaluation of Proactive, Reactive and Hybrid Routing Protocols With Mobility Model in MANETs", *International Journal of Research in Engineering and Technology (IJRET)*, Vol. 2, No. 8, pp. 254-259.

[31] Abdul Karim Abed, Gurcu Oz, and Isik Aybay, (2014) "Influence of mobility models on the performance of data dissemination and routing in wireless mobile ad hoc networks", *Computers and Electrical Engineering*, Vol. 40, No. 2, pp. 319-329.

[32] Indrani Das, D.K Lobiyal, and C.P Katti, (2014) "Effect of Node Mobility on AOMDV Protocol in MANET", *International Journal of Wireless & Mobile Networks (IJWMN)*, Vol. 6, No. 3, pp. 91-99.

[33] Karmveer Singh and Vidhi Sharma, (2014) "Performance Analysis of MANET with Reactive and Proactive Routing Protocols and Mobility Models", *International Journal for Research in Applied Science and Engineering Technology (IJRASET)*, Vol. 2, No. 5, pp. 304-312.

[34] Mohamed Wahed, Hassan Al-Mahdi, Tarek M Mahmoud, and Hassan Shaban, (2014) "The Effect of Mobility Models and Traffic Patterns on the Performance of Routing Protocols in MANETs", *International Journal of Computer Applications*, Vol. 101, No. 9, pp. 52-58.

[35] Ramanpreet Kaur and Rakesh Kumar, (2014) "Comparison Study of Routing Protocol by Varying Mobility and Traffic (CBR, VBR and TCP) Using Random Walk & Random Way Point Models", *International Journal of Engineering Trends and Technology (IJETT)*, Vol. 7, No. 4, pp. 177-183.

[36] Vasudha Sharma and Sanjeev Khambra, (2014) "Performance Comparison of DSR, OLSR and TORA Routing Protocols", *International Journal of Scientific & Technology Research*, Vol. 3, No. 8, pp. 411-415.

[37] NS3, Network Simulator 3, http://www.nsnam.org.

A Review on Cooperative Communication Protocols in Wireless World

Juhi Garg[1], Priyanka Mehta[2] and Kapil Gupta[3]

[1]Department of Electronics and Communication Engineering, FET-MITS University, Lakshmangarh, Sikar, India
`joohigrg@gmail.com`
[2]Department of Electronics and Communication Engineering, FET-MITS University, Lakshmangarh, Sikar, India
`primehta04@gmail.com`
[3]Department of Electronics and Communication Engineering, FET-MITS University, Lakshmangarh, Sikar, India
`kapil_mbm@yahoo.com`

ABSTRACT

Future generations of cellular communications requires higher data rates and a more reliable transmission link with the growth of multimedia services, while keeping satisfactory quality of service, . MIMO antenna systems have been considered as an efficient approach to address these demands by offering significant multiplexing and diversity gains over single antenna systems without increasing bandwidth and power. Although MIMO systems can unfold their huge benefit in cellular base stations, but they may face limitations when it comes to their deployment in mobile handsets.

To overcome this drawback, relays (fixed or mobile terminals) can cooperate to improve the overall system performance in cellular networks. Cooperative communications can efficiently combat the severity of fading and shadowing through the assistance of relays. It has been found that using relays the capacity and coverage of cellular networks can be extended without increasing mobile transmit power or demanding extra bandwidth.

KEYWORDS

Cooperative Communication, Coded Cooperation, MIMO, Amplify and Forward, Decode and Forward

1. INTRODUCTION

The increasing numbers of users demanding service have encouraged intensive research in wireless communications. The problem with the cooperative communications is the unreliable medium through which the signal has to travel. To mitigate the effects of wireless channel, the idea of diversity has been deployed in many wireless systems [1-3]. Diversity is a communication technique where the transmitted signal travels through various independent paths and thus the probability that all the wireless paths are in fade is made negligible. Frequency diversity, time diversity and space diversity are the three basic techniques for providing diversity to the wireless communication systems.

Multiple-input multiple-output (MIMO) systems, where the transmitters as well as receivers are equipped with multiple antennas, proved to be a breakthrough in wireless communication system which offered new degree of freedom, in spatial domain, to wireless communications. After that, MIMO became part of many modern wireless communications standards like LTE

Advanced [1], WiMAX [2], [3] and Wireless LAN [4]. However, use of MIMO in small size nodes, like used in wireless cellular networks proved to be a challenge. To address this challenge, idea of cooperative communications came into existence to implement the idea of MIMO in distributed manner. This concept says that transmitting users share each others' antennas to give a virtual MIMO concept. Though, the idea of cooperative communication was given in 2003 by Sendonaris et al. [5], [6], it is still considered an extensive research which is going to exploit its benefits in the next generation communication systems [7][8].

In this article, we discuss a detailed overview of potential benefits offered by this new trend in communication system.

2. WIRELESS COMMUNICATIONS AND MIMO SYSTEMS

Wireless communication systems have grown successfully over the last few decades and are expected to continue in the future. There is an increasing demand for high data rates today in order to support high speed interactive internet services and advanced multimedia applications such as mobile TV, online gaming etc. This trend is enormously evolving in the 4G systems.

However, transmission over wireless channel of high rate (i.e. bandwidth) faces fundamental limitations due to impairments inflicted by wireless channel such as, path loss, shadowing and fading effects. These impairments can be compensated by various ways such as by increasing transmit power, bandwidth, and/or applying powerful error control coding (ECC). However, power and bandwidth are very scarce and expensive radio resources whereas ECC yields reduced transmission rate. Thus, acquiring a high data rate together with reliable transmission over error-prone wireless channels is a major challenge for wireless system designer.

Another way to cope with the impairments offered by the wireless channel is the use of MIMO systems. In MIMO systems, multiple antennas are used at the transceiver. This arrangement can significantly increase data rate and reliability of the wireless link. MIMO systems use either VBLAST (Vertical Bell Laboratories Layered Space-Time) or DBLAST (Diagonal Bell Labs Layered Space-Time) algorithm. However, using multiple co-located antennas causes degradation in the system Quality of service (QoS) due to correlation between them. Also, due to size, cost, or hardware limitations, small handheld wireless devices may not be able to support multiple antennas [10].

To overcome the above drawback, an innovative approach known as cooperative communication has been suggested to exploit MIMO's benefit in a distributed manner. Such a technique is also called a virtual MIMO, since it allows single antenna mobile terminals to reap some of the benefits of MIMO systems. This concept is illustrated in Fig-1 below [9][10].

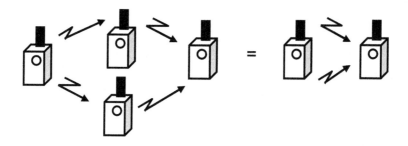

Figure 1. Illustration of MIMO and virtual MIMO systems [10]

3. WORKING PRINCIPLE

The idea of cooperation was presented by van der Meulen in 1971, which established foundation of relay channel. Cooperative communication takes advantage of broadcast nature of the wireless medium where the neighbouring nodes overhear the source's signals and relay the information to the destination. Thus, idea of creating a VAA came into existence [11].

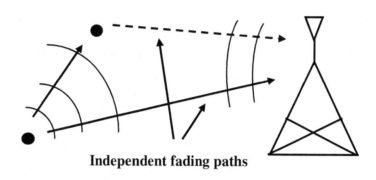

Independent fading paths

Figure 2. Cooperative Communication

4. HISTORICAL BACKGROUND

The scheme of cooperative communications is based on simple three terminal relay channel introduced by Van der Meulen [12][13]. In this, it was shown that both the transmitter and receiver nodes can be aided by a relay node to improve the rate region of the transmitter. Later, idea of cooperative communications was given by Sendonaris [5][6]. The authors prove that cooperative communications improves the achievable rate for both users, based on two user's cooperative communication scenario.

This is further extended by Lane man et al. [14-16]. The authors introduced the idea of amplify-and-forward, decode-and-forward, and dynamic decode-and-forward protocol. The outage probability is used as a performance metric to advocate the idea of cooperative communication.

The concept of coded cooperation was given by Hunter [17], [18] which is given as the integration of cooperative communication and channel coding. The users code their information into block composed of two parts. The transmitting user transmits the first part of block and second part is transmitted by another cooperating user if it receives the first part successfully, otherwise, user itself transmits the second part of the block. In this way, diversity gain is exploited to efficiently counter the wireless channel impairment.

Later, various channel coding schemes have been applied to this scenario. LDPC [19] was first applied by Khojastepour et al. [20]. Turbo codes et al. [21] was investigated by various researchers [22-24]. Extensive work has been implemented by Hunter et al. [17], [18] where they have utilized convolution codes to cooperative communication.

Thus, in the wireless transmission, the transmission of a node is physically broadcasted. Hence, even though a node is not the next-hop node of a transmitted data, it can overhear it. In the conventional transmission, only the next-hop node of the received data accepts it and all other nodes just ignore it. We call this conventional transmission a direct transmission in this paper.

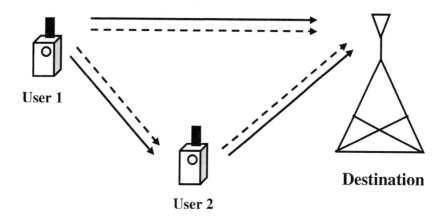

Figure 3. In Cooperative Communication each mobile is both a relay and a user

However, if we appropriately utilize the broadcasting nature of the wireless transmission, we can improve the performance of transmission such as its reliability and throughput. The cooperative transmission uses this broadcasting nature of wireless transmission to improve the performance of the transmission. In the cooperative transmission, if a node overhears a data transmission, it may forward it to its next-hop node after performing some processing that depends on cooperative transmission scheme. We call this node a cooperative node. Hence, the next-hop node of the data may receive multiple data from the original transmitter node and from several cooperative nodes and by appropriately combining and decoding them, the performance of the transmission could be improved compared with that of the direct transmission.

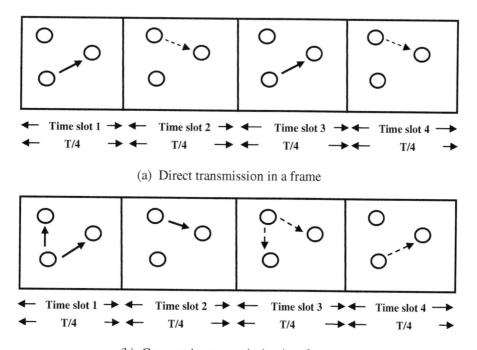

(a) Direct transmission in a frame

(b) Cooperative transmission in a frame

Figure 4. Direct and Cooperative Communication in a frame

5. PREVIOUS WORK

Cooperative relaying is an effective technique to combat multipath fading and enhance system capacity and coverage. This section provides a literature survey on various aspects of cooperative relaying.

Initially, capacity analyses of Gaussian relay channel was comprehensively studied in [25]. In [26], various relaying strategies for wireless networks have been studied. End-to-end error and outage performance of non-regenerative relay network have been studied in [27]. Here, all nodes are equipped with single antenna which operates in Nakagami-m fading channel. Outage probability of multi-hop amplifies and forward relay systems have been analyzed in [28]. In [29], closed form expressions of average bit error rate (ABER), amount of fading and outage probability (OP) have been derived for Rayleigh-fading channel where relay works in non-regenerative mode and destination performs equal gain combining (EGC). In [30] and [31], closed-form expressions are derived for the moment generating function (MGF), probability density function (PDF), and cumulative distribution function (CDF) of the statistically independent Gamma random variables. An amplify-and-forward (AF) relay network has been studied in [32] where a source communicates with the user having best channel conditions through an intermediate relay that serves to multiple users. Spatial reuse of relay time-slot in interference limited channel has been analyzed in [33] for AF cooperative relaying. In [34], - Laneman et al. has developed and analyzed distributed space-time coded cooperative diversity protocols for improving spectral efficiency. Error performance has been analyzed in [35] when cooperative relay system operates in asymmetric channels. In [36], outage performance of MIMO relay channel has been investigated when source and relay use same orthogonal space time block code. Selection cooperation in network scenario has been discussed in [37] where multiple sources transmit their message and support each other to forward signals to their respective destinations. End-to end error performance of selection cooperation has been presented in [38]. In [39], optimum power allocation for multi-antenna relay network at the expense of increased computational complexity has been presented.

An overview of various cooperation schemes related to their implementation has been discussed in [40]. In [41], closed form expressions of OP and bit error rate (BER) for binary phase shift keying (BPSK) are derived for case where communication between source and destination is supported by multiple-antenna relay and both relay and destination performs MRC of signals in Rayleigh fading channel. OP and average error rate of two-hop multi-antenna relay based system for the case when relay performs selection combining (SC) of signals and destination performs MRC of signals are analyzed in [42] and [43], respectively. In [44], closed form expressions for OP and BER have been derived when multi-antenna relay network operates in correlated Nakagami-m fading channel and both the relay as well as destination performs MRC of signal. Closed form expression of outage probability has been derived in [45]; here, communication between source and destination is supported by two multi-antenna relay nodes. A new efficient scheme for cooperative wireless networking based on linear network codes has been discussed in [46].

6. BASIC RELAYING PROTOCOLS

In cooperative communications, the transmitting user not only broadcast their own message but they also relay information, on behalf of each other, to the destination. The strategy, by which the information is relayed to the destination, is known as protocol. Various protocols have been introduced so far. Here, we describe some of the basic relaying protocols.

6.1. Amplify and Forward

This is the simplest protocol. Here, the information received by a user from original transmitter is amplified and then forwarded to the destination. Based on the principle of amplifying repeaters [47], amplify-and-forward protocol was formally introduced by Lane man et al. [48]

6.2. Decode and Forward

In this relaying protocol the partner users decodes the message received from original transmitter, re-encodes and then forwards it to the destination. Thomas M. Cover and Abbas A. El Gamal are considered as its pioneer [50] and later the idea was further explored by many authors with the name of Decode-and-Forward [16], [15], [50].

6.3. Compress and Forward

In this relaying protocol, the message is decoded from the transmitter and the partner user forwards a compressed version of it to the destination, so as to get the diversity benefits [49].

7. COOPERATIVE RELAYING TECHNIQUES

In this section, we will discuss Different Cooperative Protocols or Transmissions Techniques used in Cooperative Communication. Cooperative communications protocols can be generally categorized into fixed relaying schemes and adaptive relaying schemes. In this section, we describe both of these schemes along with both single relay and multi relay scenario.

7.1. Cooperation Protocols

A cooperation strategy is modelled into two orthogonal phases, to avoid interference between the two phases, either in TDMA or FDMA.

In phase 1, source broadcast information to its destination, and the information is also received by the relay (due to broadcast) at the same time as shown in figure 5 below.

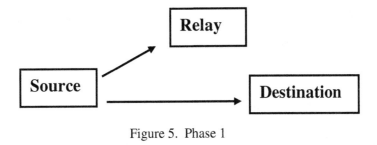

Figure 5. Phase 1

In phase 2, the relay forwards or retransmits the information to the destination as shown in figure 6 below.

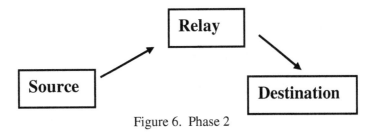

Figure 6. Phase 2

Fig- 7 below depicts a general relay channel, where the source transmits with power P1 and the relay transmits with power P2. Here, we will consider the special case when both source and relay transmit with equal power P. Optimal power allocation is a vast topic so can be consider for future work.

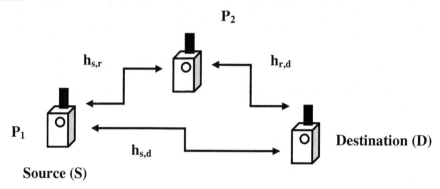

Figure 7. Simplified Cooperation Model [51]

In phase 1, source forwards its information to both the destination and the relay. The received signals Y_{sd} at the destination and Y_{sr} at the relay, can be written as

$$Y_{sd} = \sqrt{P}h_{sd}x + n_{sd} \quad (1)$$
$$Y_{sr} = \sqrt{P}h_{sr}x + n_{sr} \quad (2)$$

where P is the transmitted power at source, x is the transmitted information symbol, and n_{sd} and n_{sr} are additive noise. In (1) and (2), h_{sd} and h_{sr} are the channel fades between the source and destination and the relay, respectively, and are modelled as Rayleigh flat fading channels. Rayleigh flat fading channel can be mathematically modelled as complex Gaussian random variable. It is given as z = x + jy, where real and imaginary parts are zero mean independent and identically distributed (i.i.d) Gaussian random variables. The noise terms n_{sd} and n_{sr} are modelled as zero-mean complex Gaussian random variables with variance No.

In phase 2, relay forwards source's signal to the destination, and this can be modelled as

$$Y_{rd} = h_{rd}q(Y_{sr}) + n_{rd} \quad (3)$$

where the function q (·) depends upon processing which is implemented at the relay node [52].

7.2. Fixed Cooperation Strategies

In fixed relaying, channel resources are divided between source and relay in a fixed (deterministic) manner. The processing at relay differs according to the employed protocols. The most common techniques are the fixed AF relaying protocol and the fixed relaying DF protocol [15][34].

7.2.1. Fixed Amplify & Forward (Single Relay)

In a fixed AF relaying protocol, or AF protocol, the relay scales the received version and transmits its amplified version to the destination. The amplify-and-forward scheme is presented in Fig-8.

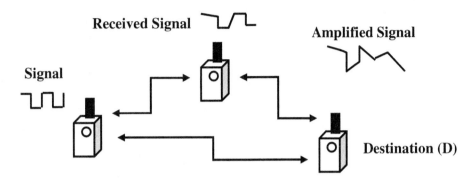

Figure 8. Amplify and Forward System Model [51]

The amplify-and-forward relay channel can be modelled as follows. The signal transmitted from the source x to both the relay and destination can be given as

$$Y_{sr} = \sqrt{P}h_{sr}x + n_{sr} \text{ and } Y_{sd} = \sqrt{P}h_{sd}x + n_{sd} \quad (4)$$

where h_{sr} and h_{sd} are channel fades between the source and the relay and destination, respectively. The terms n_{sr} and n_{sd} denote the additive white Gaussian noise with zero-mean and variance N_0. Here, the relay amplifies the source signal and forwards it to the destination, to equalize the effect of the channel fades between the source and the relay. The relay scales the received signal by a factor that is inversely proportional to the received power, which is denoted by

$$\beta = \frac{\sqrt{P}}{\sqrt{P|h_{sr}| + N_o}} \quad (5)$$

The signal transmitted from relay is therefore, given by βY_{sr} and has power P equal to the power of the signal transmitted from the source. In phase 2, the relay amplifies the received signal from source and forwards it to the destination.

The received signal, in phase 2, at the destination, according to (4) is given as

$$Y_{rd} = \frac{\sqrt{P}}{\sqrt{P|h_{sr}| + N_o}} h_{rd}Y_{sr} + n_{rd} \quad (6)$$

Here h_{rd} is the channel coefficient from relay to the destination and n_{sd} is an additive noise.

7.2.2. Fixed Amplify & Forward (Multi Relay)

An amplify-and-forward protocol does not suffer from any error propagation problem because the relay does not need to perform any hard-decision operation on the received signal. First we discuss the multi-node amplify-and-forward protocol and then analyze its two relaying strategies. In the first phase, each relay forwards the source's signal to the destination, while in the second phase each relay forwards a combined signal from source and previous relays. Fig-9 shows the multi-node Amplify and Forward system model.

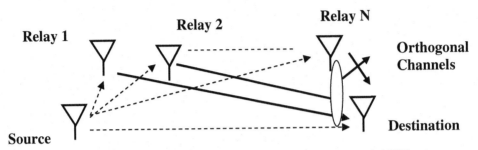

Figure 9. Multi Node Amplify and Forward System Model [52]

7.2.3. Fixed Decode and Forward (Single Relay)

Another possibility at the relay node is to decode the received signal at the relay, re-encode it, and then retransmit it to the receiver. The decode-and-forward scheme is presented in Fig-10. This kind of relaying is often called as DF scheme. If the decoded signal at the relay is denoted by x', the transmitted signal from the relay can be denoted by x', given that x' has unit variance.

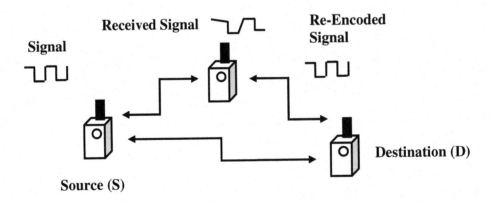

Figure 10. Decode and Forward System Model [52]

If an incorrect signal is forwarded to the destination, then decoding at the destination is meaningless. It is clear that the diversity achieved for such a scheme is only one, because performance of the system is limited by the worst link from the source–relay and source–destination.

Although fixed DF relaying has the advantage of reducing the effects of additive noise at the relay over AF relaying, it forwards erroneously detected signals to the destination, causing error propagation that can diminish the performance of the system. Mutual information between the source and destination is limited by the mutual information of the weakest link between the source–relay and the combined channel from the source–destination and relay–destination. The received signal at the destination in Phase 2 can be given as

$$Y_{rd} = \sqrt{\beta_2} h_{rd} x + n_{rd} \qquad (7)$$

With the help of channel coefficients h_{sd} (between source and destination) and h_{rd} (between relay and destination), the destination detects the transmitted symbols by jointly combining the received signal Y_{sd} from the source in eq. (1) and Y_{rd} from the relay in eq. (7).

7.2.4. Fixed Decode and Forward (Multi Relay)

Here, we consider a N-relay (2 Relay For Simulation Purpose) wireless network, where information is to be transmitted from a source to a destination. Due to the broadcast nature of the wireless channel, relays overhear the transmitted information and thus cooperate with the source to send its data. The wireless link between any two nodes is modelled as a Rayleigh fading channel with AWGN.

The channel fades for different links are assumed to be statistically independent because the relays are usually spatially separated. The additive noise at receiving terminals is modelled as zero-mean, complex Gaussian random variables with variance N_o. The relays are assumed to transmit over orthogonal channels, thus there is no inter-relay interference.

In DF protocol cooperation strategy, each relay can combine the signal received from the source along with the signals transmitted by previous relays, decode it and then retransmit it to the receiver after re-encoding it again. A general cooperation scenario, denoted as C(m) ($1 \leq m \leq N - 1$), is implemented in which each relay combines the signals received from m previous relays along with that received from the source.

The multi relay decode and forward scenario is shown in Fig- 11, in which each relay combines the signals received from previous relays along with that received from the source. In all considered cooperation scenarios, destination coherently combines the signals received from the source and relays. Now, we are presenting the system model for a general cooperative scheme C(m) for any $1 \leq m \leq N - 1$, where each relay decodes the information after combining the signals received from the source and previous m relays. We consider 2 relays for simulation purpose.

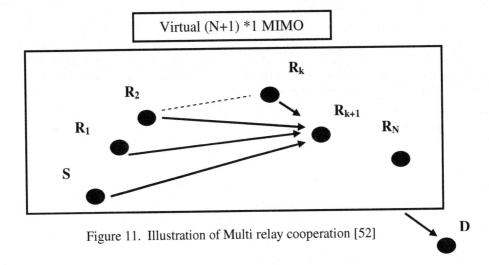

Figure 11. Illustration of Multi relay cooperation [52]

In General Multi Node cooperation protocol has (N+1) phases as stated in AF section but three in our case. In phase 1, source transmits information to both destination and the ith (1st) relay, which can be modelled as,

$$Y_{sd} = \sqrt{P_o} h_{sd} x + n_{sd} \tag{8}$$

$$Y_{s,ri} = \sqrt{P_o} h_{sri} x + n_{sri} \quad 1 \leq i \leq N \tag{9}$$

where P_o is the power transmitted at the source, x is the transmitted symbol, h_{sd} and h_{sri} are the fading coefficients between source and destination, and i^{th} relay, respectively. The terms n_{sd} and n_{sri} denote the AWGN channel noise. In phase 2, the 1st relay decodes the signal it receives from source, re-encode & send it to other relay & the destination. Second relay combines the received signals from the 1st relay and source as follow

$$Y_{r2} = \sqrt{P_o} h^*_{s,r2} Y_{s,r2} + \sqrt{P_i} h^*_{r1,r2} Y_{r1,r2} \tag{10}$$

where $h_{r1,r2}$ is the channel fading coefficient between the 1st and the 2nd relay, $Y_{r1,r2}$ denotes the signal received at 2nd relay from the 1st relay, which is given as

$$Y_{r1,r2} = \sqrt{P_i} h_{r1,r2} x + n_{r1,r2} \tag{11}$$

where P_i is the power transmitted at relay 1, finally in phase (N + 1), the destination coherently combines all of the received signals, which are modelled as

$$Y_d = \sqrt{P_o} h^*_{s,d} Y_{sd} + \sum_{i=1}^{N} \sqrt{P_i} h^*_{ri,d} Y_{ri,d} \tag{12}$$

7.3. Other Cooperation Strategies

Besides the techniques for fixed relaying, there are other techniques, such as compress-and forward cooperation and coded cooperation. Here, we discuss an overview of each of them.

7.3.1. Compress and Forward

The main difference between compress-and forward and decode/amplify-and-forward is that in the later the relay transmits a copy of the received message, while in compress and forward the relay transmits a quantized and compressed version of the received message. Therefore, the destination performs the reception function by combining the received message from both source node and its compressed version from the relay node.

The quantization and compression process at relay node is a process of source encoding, i.e., representation of each received message as a sequence of symbols. Let us assume that these symbols are binary digits (bits). At the destination, an estimate of the compressed message is obtained by decoding the received sequence of bits. This decoding operation involves mapping of received bits into a set of values that estimate the transmitted message. This mapping process normally involves the introduction of distortion, which can be considered as a form of attenuation and noise [25], [53].

7.3.2. Coded Cooperation

Coded cooperation differs from the previous schemes such that the cooperation is implemented at the level of the channel coding subsystem. We know in both amplify and forward and decode-and-forward schemes, the relay repeats the bits sent by the source. In coded cooperation incremental redundancy at relay, which when combined at the receiver with the codeword sent by the source, results in a codeword with larger redundancy [54].

7.4. Adaptive Cooperation Strategies

Fixed relaying has advantage of easy implementation and disadvantage of low bandwidth efficiency. This is because half of the channel resources are allocated to relay for transmission, which leads to reduced overall rate, because in such a scenario, a high percentage of the packets transmitted by the source to the destination could be received correctly by the destination and the relay's transmissions would be wasted.

To overcome this problem, adaptive relaying protocols are developed to improve the inefficiency. Here, we will discuss two strategies: selective DF relaying and incremental relaying [52].

7.4.1. Selective DF Relaying

In this relaying scheme, the relay decodes the received signal and forwards decoded information to the destination, if the signal-to noise ratio of a signal at the relay exceeds a certain threshold. On the other hand, if the channel between the source and the relay suffers, such that the signal-to-noise ratio falls below the threshold, the relay remains idle. Selective relaying shows improvement upon the performance of fixed DF relaying, as the threshold at the relay can be used to overcome the inherent problem in fixed DF relaying in which the relay forwards all decoded signals to the destination although some signals are incorrect. We knew, if the SNR in source–relay link exceeds the threshold, the relay is able to decode source's signal correctly. In this case, SNR of the combined MRC signal at destination is given as sum of the received SNR from the source and the relay.

7.4.2. Incremental Relaying

In incremental relaying, there is a feedback channel from the destination to the relay. The destination sends an acknowledgement to the relay if it was able to receive the source's message correctly. This protocol has the best spectral efficiency among the previously described protocols because the relay does not always need to transmit, and hence second transmission phase becomes opportunistic depending on the channel state condition of channel between the source and the destination. Nevertheless, incremental relaying achieves a diversity order of two as mentioned below:

In incremental relaying, if the source transmission in the first phase was successful, then there is no second phase and source transmits new information in the next time slot. On the other hand, if source transmission was not successful in the first phase, then relay can use any of the fixed relaying protocols to transmit the source signal from the first phase. The transmission rate (R) is random in incremental relaying. If the first phase was successful, the transmission rate is R, while if it was in outage, the transmission rate becomes R/2 as in fixed relaying.

8. APPLICATIONS

This section highlights some of the areas where the cooperative relaying strategies can be applied.

8.1. Virtual antenna array

The field of high-data-rate, spectrally efficient and reliable wireless communication, is currently receiving much attention. It is a well known fact that the use of MIMO antenna system improves the diversity gain of wireless systems. However, multi-antenna technique is not attractive for tiny wireless nodes due to limited hardware and signal processing capability. Diversity can be achieved through user cooperation, whereby mobile users share their physical resources to create a virtual array, which thus, removes the burden of multiple antennas on wireless terminals.

8.2. Wireless ad-hoc network

Ad hoc network is a self organizing network without any centralized infrastructure. In this n/w, distributed nodes form a temporary functional network and support seamless leaving or joining of nodes. Such network has been deployed for military communication and civilian applications including commercial and educational use etc.

8.3. Wireless sensor network

Lifetime of sensor network can be increased by deploying cooperative relaying hence, energy consumption in sensor nodes got reduced. We knew, communication through weaker channels requires huge energy as compared to relatively strong channels. So, careful incorporation of cooperating relay nodes into routing process can select better communication links and precious battery power can be saved.

8.4. Cooperative sensing for cognitive radio

In cognitive radio system, secondary users can utilize the resources which are employed for licensed primary users. When primary users want to use their licensed resources, secondary users have to vacant these resources. Thus, secondary users have to constantly sense the presence of primary user. Probability of false alarming can be reduced with the help of spatially distributed nodes, which thus improve the channel sensing reliability by sharing the information [55].

9. TRACKS FOR FUTURE WORK

This paper investigated the performance of some of the basic cooperative relaying schemes. For practical implementation of such schemes in a network environment, it is necessary to investigate several other issues, some of which are briefly outlined below:

9.1. System modelling taking inference into consideration

In a network scenario, cooperative relay system suffers from co-channel and adjacent channel interference. Thus, the effect of the interference needs to be analysed.

9.2. Power optimization based on link condition

Wireless nodes generally have limited battery power. Power can be allocated if relay based systems have some feedback mechanism, based on link condition [34]. Such dynamic allocation may save battery power or boost the data transfer rate and hence the optimization area to be investigated.

9.3. Full duplex operation of relays

Relay operating in half-duplex mode creates a wide system bandwidth expansion. Full-duplex relay operating in single frequency are used to solve this problem. Thus, the effect of full-duplex relay operation needs to be investigated.

9.4. Complexity performance trade off

Relays can process the signal in non-regenerative or regenerative mode depending on their functionality. Non-regenerative mode puts less processing burden on the relay as compared to regenerative mode of operation; hence, it is often preferred when complexity and/ or latency are needed to be analysed. Scope has been found to exist in future for non-regenerative mode of relay operation. Noise amplification is a major issue in this operation.

9.5. Modelling of relay link in various environments

Wireless propagation suffers with path loss, shadowing and fading which depends on regional - geography. Statistical behaviours of various propagation environments are available in literatures. Relay link can be easily modelled in such propagation environment.

9.6. Relay selection

In the present state of art, wireless relay nodes are half-duplex in nature. Participation of large number of relay increase the diversity order, but spectrum efficiency of system may suffer. Therefore, selection of the nodes and optimum number of nodes which should be selected to optimize the performance in terms of rate and reliability is an interesting field of research.

9.7. Spectrum efficiency

Orthogonal transmission from relay to receiver in TDMA mode can affect the spectrum efficiency. Relay can interact with the receiver in CDMA mode which avoids bandwidth expansion. Therefore, orthogonal code design for distributed relay node may be possible area for research.

9.8. Node cooperation with cross-Layer design

Node cooperation is an efficient technique to improve the performance of WSN's. Sensor nodes are powered by small battery that cannot be easily recharged. So reducing energy consumption is a wide issue for such type of network. Protocol layering provides modularity that facilitates standardization and implementation. Unfortunately, layering precludes the benefits of joint optimization across protocol layers; hence, precious battery power cannot be efficiently utilized. In this regard, it would be interesting to investigate cross-layer optimization [56].

9.9. Bidirectional user cooperation

In this work, we are assuming that relay terminals do not have their own data and they are just forwarding data received from source. In user cooperation, user's terminal not only transmits its own data, but also relays other user's data by sharing some of the resources. In this regard, it would be interesting to investigate bidirectional user cooperation extension.

9.10. Base station cooperation

In single cellular system, user terminals can communicate with parent base station. Users near the outskirts of cell can communicate with neighbouring base stations and thus becomes a source for generating interfering signals. To overcome this, neighbouring base stations can also cooperate with parent base station and perform joint decoding of received signals. The work presented for relay cooperation can be extended for base station cooperation.

10. CHALLENGES

Helping out users in a cooperative fashion has its price. Here, we will describe the challenges that incur in systems with cooperative communication incorporated [57][58].

10.1. Complex Schedulers

Relaying requires more sophisticated schedulers since not only traffic of different users and applications needs to be scheduled but also the relayed data flows. This function decides how many resources are scheduled for a single user (or relay node). This function affects the achieved throughput of the system. Practical implementations of a scheduler also consider Automatic Repeat-Request (ARQ) protocols and Quality-of-Service (QOS) classes, which experience different priorities in the scheduling process.

10.2. Increased Overhead

A full system functioning requires handovers, synchronization, extra security, etc. This clearly induces an increased overhead w.r.t to a system that does not use relaying.

10.3. Increased Interference

If the offered power savings are not used to decrease the transmission power of the relay nodes but rather to boost capacity or coverage, then relaying will certainly generate extra intra and inter-cell interference, which potentially causes the system performance to deteriorate. So cooperative relying is more suitable for 3G/4G systems which are more tolerant to interference. Many Interference mitigation schemes have been proposed for cooperative communication e.g., successive interference cancellation (SIC) scheme in wireless communication network. Zero forcing (ZF) or minimum mean square error (MMSE) receivers are used to mitigate co-channel interference (CCI) in the interference cancellation strategy. For efficient utilization of scarce radio spectrum and codes, a centralized medium access control (MAC) protocol is proposed to coordinate the code assignment and channel access among the users and relays [59], [60].

10.4. Increased End-To-End Latency

Relaying typically involves the reception and decoding of the entire data packet before it can be retransmitted. If delay-sensitive services are being supported, such as voice or the increasingly popular multimedia web services, then the latency induced by the decoding may become detrimental. Latency increases with the number of relays and also with the use of interleavers, such as utilized in GSM voice traffic. To circumvent this latency, either simple transparent relaying (i.e. AF relaying) or some advanced decoding methods need to be used.

10.5. More Channel Estimates

The use of relays effectively increases the number of wireless channels. This requires the estimation of more channel coefficients and hence more pilot symbols need to be provided if coherent modulation was to be used.

11. CONCLUSIONS

As stated at the outset, the field of high-data-rate, efficient and reliable wireless communication, is currently receiving much attention. Cooperative transmission is emerging as an effective technique for combating effects of path loss, shadowing, and multi-path fading. This tutorial elaborates wireless cooperative communication, a technique that both allow single antenna mobiles to share their antennas and reap the benefits of multiple antenna systems. Cooperative relaying provides diversity gain, reduces outage probability and improves BER performance. Various types of relays, mode of operation, applications, and tracks for future work have been discussed here. This paper will be helpful for incorporating relay based system in real scenario. Throughout this paper we focused on only two Protocols viz AF & DF but there are many other protocols as well that deserve attention. We also didn't include Power Allocation i.e. at what power; source/relay should transmit without causing interference for others.

REFERENCES

[1] 3GPP TR 36.814 V1.2.1, "*Further advancements for EUTRA: Physical layer aspects,*" Technical Specification Group Radio Access n/w, June 2009.

[2] I. P802.16j/D9, "Draft amendment to IEEE standard for local and metropolitan area network part 16: Air interface for fixed and mobile broadband wireless access systems: Multihop relay specification," May 2009.

[3] Yang Yang; Honglin Hu; Jing Xu; Guoqiang Mao, "Relay technologies for WiMAX and LTE-Advanced mobile systems," *IEEE Comm. Magazine*, vol. 47, no. 10, pp. 100–105, October 2009.

[4] "IEEE draft standard for local and metropolitan area networks; part 11: *Wireless LAN medium access control (MAC) and physical layer (PHY) specification: amendment: enhancements for higher through- put,* IEEE Draft Std.802.11n (d2)," 2007.

[5] A. Sendonaris, E. Erkip and B. Aazhang, "User cooperation diversity- Part I: System description," *IEEE Transaction on Communications*, vol. 51, no. 11, pp. 1927–1938, Nov. 2003.

[6] A. Sendonaris, E. Erkip and B. Aazhang, "User cooperation diversity-Part II: Implementation aspects and performance analysis," *IEEE Trans. on Commun.*, vol. 51, no. 11, pp. 1939–1948, Nov. 2003.

[7] L. Sankar, G. Kramer and N. B. Mandayam, "Dedicated-relay vs. user cooperation in time-duplexed multi-access networks," *Journal of Communications*, vol. 6, no. 9, pp. 330–339, July 2011.

[8] Hongzheng Wang; Shaodan Ma; T. S. Ng, "On performance of cooperative communication systems with spatial random relays," *IEEE Trans. on Commun.*, vol. 59, no. 4, pp. 1190–1199, 2011.

[9] Birsen, Sirkeci, "*Distributed Distributed Cooperative Communication in Large-Scale Wireless Networks,*" Cornell University 2006.

[10] Quang Trung Duong, "On Cooperative Communications and Its Application to Mobile Multimedia," Blekinge Institute of Technology, ISBN: 978-91-7295-167-9, 2010.

[11] Y.W. Peter Hong, Wan-Jen Huang & C.C. Jay Kuo, "*Cooperative communications and Networking: Technologies and System Design,*" Springer, (2010) ISBN: 1441971939.

[12] E. C. Van Der Meulen, "Transmission of information in a T-terminal discrete memory less channel," Ph.D. Disser., Dept. of Statistics, University of California, Berkeley, 1968.

[13] Edward C. Van Der Meulen, "Three terminal communication channels," *Adv. Appl. Prob.*, vol. 3, pp. 120–154, 1971.

[14] J. N. Laneman and G. W. Worn ell, "Energy-efficient antenna sharing and relaying for wireless networks," *in IEEE Wireless Commun. and Networking Conference* (WCNC-2000), vol. 1, pp. 7–12, Sep 2000.

[15] J. N. Laneman, D. N. C. Tse and G. W. Wornell, "Cooperative diversity in wireless networks: efficient protocols and outage behaviour," *IEEE Trans. on Information Theory*, vol. 50, no. 12, pp. 3062–3080, Dec. 2004.

[16] J. N. Laneman, "Cooperative diversity in wireless networks: Algorithms and architectures," Ph.D. dissertation, Massachusetts Institute of Technology, Cambridge, MA, Aug 2002.

[17] T. E. Hunter and A. Nosratinia, "Co-operation diversity through coding," *IEEE International Symposium on Information Theory*, pp. 220, July 2002.

[18] T. E. Hunter and A. Nosratinia, "Diversity through coded co-operation," *IEEE Transactions on Wireless Commun.*, vol. 5, no. 2, pp. 283–289, Feb. 2006.

[19] R. G. Gallager, "Low density parity-check code," Ph.D. Dissertation, Massachusetts Institute of Technology, Cambridge, MA, 1963.

[20] M. A. Khojastepour, N. Ahmed and B. Aazhang, "Code design for the relay channel and factor graph decoding," *Conference Record of the Thirty-Eighth Asilomar Conf. On Signals, Systems and Computers*, vol. 2, pp. 2000–2004, 2004.

[21] C. Berrou, A. Glavieux and P. Thitimajshima, "Near Shannon limit error-correcting coding and decoding: Turbo-codes," *IEEE International Conference on Communications*, vol. 2, pp. 1064–1070, 1993.

[22] Z. Zhang and T. M. Duman, "Capacity-approaching turbo coding and iterative decoding for relay channels," *IEEE Trans. on Commun.*, vol. 53, no. 11, pp. 1895–1905, Nov. 2005.

[23] B. Zhao and M. C. Valenti, "Distributed turbo-coded diversity for the relay channel," *IEEE Electronics Letters*, vol. 39, no. 10, pp. 786–787, May 2003.

[24] Bin Zhao and Matthew C. Valenti, "*Cooperative diversity using distributed turbo codes,*" Virginia Technical Symposium on Wireless Personal Communications, (Blacksburg, VA), June 2003.

[25] T. Cover and A. E. Gamal, "Capacity theorems for the relay channel," *IEEE Trans. on Information Theory*, vol. 25, no. 5, pp. 572-84, Sep. 1979.

[26] P. Herhold, E. Zimmer-mann and G. Fettweis, "*Cooperative multi-hop transmission in wireless networks,*" Computer N/w, Vol. 49, no. 3, pp. 299-324, 2005. Selected Paper from the European Wireless 2004 Conference.

[27] M. S. Alouini and M. O. Hasna, "End-to-end performance of transmission systems with relays over Rayleigh-fading channels", *IEEE Trans. on Wireless Commun.*, vol. 2, no. 6, pp. 1126-31, Nov. 2003.

[28] C. Conne and Il-Min Kim, "Outage probability of multi-hop amplify-and-forward relay systems," *IEEE Trans. on Wireless Communications*, vol. 9, no. 3, pp. 1139-49, Mar. 2010.

[29] S. Ikki and M. H. Ahmed, "Performance analysis of cooperative diversity using equal gain combining (EGC) technique over Rayleigh fading channels," *IEEE International Conf. on Commun.*, pp. 5336-41, Jun. 2007.

[30] G. K. Karagiannidis, T. A. Tsiftsis, R. K. Mallik, N. C. Sagias and S. A. Kotsopoulos, "Closed-form bounds for multi-hop relayed communications in Nakagami-m fading", *IEEE International Conf. on Communications*, vol. 4, pp. 2362-6, May 2005.

[31] G. K. Karagiannidis, T. A. Tsiftsis and R. K. Mallik, "Bounds for multi-hop relayed communications in Nakagami-m fading," *IEEE Trans. on Communications*, vol. 54, no.1, pp. 18-22, Jan. 2006.

[32] J. B. Kim and D. Kim, "Comparison of two SNR-based feedback schemes in multi-user dual-hop amplify and forward relaying networks", IEEE *Commun. Letters,* vol. 12, no.8, pp. 557- 9, Aug. 2008.

[33] A. Agustin and J. Vidal, "Amplify-and-forward cooperation under interference-limited spatial reuse of the relay slot", *IEEE Trans. on Wireless Commun.*, vol. 7, no. 5, pp. 1952-62, May 2008.

[34] J. N. Laneman and G. W. Wornell, "Distributed space-time-coded protocols for exploiting cooperative diversity in wireless networks", *IEEE Trans. on Information Theory*, vol. 49, no. 10, pp. 2415-25, Oct. 2003.

[35] A. Adinoyi and H. Yanikomeroglu, "On the performance of co-operative wireless fixed relay in asymmetric channels," *IEEE Global Telecommun. Conf.*, pp. 1-5, Dec. 2006.

[36] L. Vandendorpe and B. K. Chalise, "Outage probability analysis of a MIMO relay channel with orthogonal space-time block codes," *IEEE Commun. Letters*, vol. 12, no. 4, pp. 280-2, Apr. 2008.

[37] E. Beres and R. Adve, "Selection cooperation in multi-source co-operative networks", *IEEE Trans. on Wireless Communications*, vol. 7, no. 1, pp. 118-27, Jan. 2008.

[38] A. Adinoyi, Y. Fan, H. Yanikomeroglu and H. V. Poor, "On the performance of selection relaying," *IEEE Vehicular Technology Conference*, pp. 1-5, Sep. 2008.

[39] A. Talebi and A. W. A. Krzymien, "Optimized power allocation for multiple-antenna multi-relay cooperative communication system," *IEEE International Conf. on Spread Spectrum Techniques and Applications*, pp. 222-226, Aug. 2008.

[40] H. Katiyar, R. Bhattacharjee and B. S. Paul, "User cooperation in TDMA wireless system," *IETE Technical Review Journal*, vol. 25, pp. 270-6, Sep. 2008.

[41] H. Katiyar and R. Bhattacharjee, "Performance of MRC combining multi-antenna co-operative relay network," *AEU - International Journal of Electronics and Communications (Elsevier)*, vol. 64, no. 10, pp. 988-91, 2010.

[42] R. Bhattacharjee and H. Katiyar, "Outage performance of two-hop multi-antenna co-operative relaying in Rayleigh fading channel," *IEEE Electronics Letters*, vol. 45, no.17, pp. 881-3, Aug. 2009.

[43] H. Katiyar and R. Bhattacharjee, "Average error rate of multi-antenna decode and forward cooperative relay network," *IEEE INDICON*, DAIICT Ahmadabad, pp. 18-20, 2009.

[44] R. Bhattacharjee and H. Katiyar, "Performance of two-hop regenerative relay network under correlated Nakagami-m fading at multi-antenna relay," *IEEE Commun. Letters*, vol. 13, no. 11, pp. 820-2, Nov. 2009.

[45] H. Katiyar and R. Bhattacharjee, "Outage performance of multi-antenna relay co-operation in the absence of direct link," *IEEE Commun. Letters*, vol. 15, no. 4, pp. 398-400, Apr. 2011.

[46] M. Xiao and M. Skoglund, "Multiple-user cooperative communications based on linear network coding," *IEEE Trans. on Commun.*, vol. 58, no. 12, pp. 3345-51, Dec 2010.

[47] B. Schein and R. Gallager, "*The Gaussian parallel relay network*," pp. 22, June 2000.

[48] J. N. Laneman, G. W. Wornell, and D. N. C. Tse, "An efficient protocol for realizing cooperative diversity in wireless networks," *IEEE Intern. Symposium Information Theory*, pp. 294, June 2001.

[49] A. E. Gamal and T. Cover, "Capacity theorems for the relay channel," *IEEE Trans. on Information Theory*, vol. IT-25, no. 5, pp. 572–584, Sep. 1979.

[50] G. D. Menghwar and C. F. Mecklenbräuker, "Cooperative versus non-cooperative communications," *in 2nd International Conference on Computer, Control and Communication*, 2009 (IC4 2009), Karachi, Pakistan, Feb. 2009.

[51] K. Alexopoulos, "Performance Analysis of Decode-and-Forward with Cooperative Diversity and Alamouti Cooperative Space-Time Coding in Clustered Multihop Wireless Networks," N.P.S Monterey California, 2008.

[52] K. J. Rayliu, A. K. Sadek, Weifengsu, and Andres Kwasinski, "Cooperative Communications and Networking," *Cambridge University Press*, ISBN-13 978-0-511-46548-2, 2009.

[53] G. Kramer, M. Gastpar and P. Gupta, "Co-operative strategies and capacity theorems for relay networks," *IEEE Trans. on Information Theory*, 51(9):3037–306, pp. 3037-3063, Sept. 2005.

[54] M. Janani, A. Hedayat, T. E. Hunter and A. Nosratinia, "Coded cooperation in wireless communications: space-time transmission and iterative decoding," *IEEE Trans. on Signal Processing*, 52:362–370, February 2004. pp. 362-371.

[55] E. Peh and Y. C. Liang, "Optimization for co-operative sensing in cognitive radio networks", *IEEE Wireless Commun. and Networking Conference*, pp. 27-32, Mar. 2007.

[56] Y. Q. Zhang, "Cross-layer design for QoS support in multi-hop wireless networks", *IEEE Trans. on Signal Processing*, vol. 96, no. 1, pp. 64-76, Jan. 2008.

[57] A. Nosratinia, T. E. Hunter and A. Hedayat, "Cooperative Communication in Wireless Networks," IEEE *Commun. Magazine*, vol. 42, no. 10, pp. 74-80, 2004.

[58] Mischa Dohler, Yonghui Li, *"Cooperative Communications Hardware, Channel & Phy,"* John Wiley & Sons, 2010 ISBN 978-0-470-99768-0

[59] Eun Cheol Kim and Jae Sang Cha, *"Successive interference cancellation for cooperative communication systems,"* International Conference on Hybrid Information Technology, 2009, pp. 652-656.

[60] Kuang-Hao Liu, Hsiang-Yi Shin and Hsiao-Hwa Chen, *"Interference-resistant cooperative wireless networks based on complementary codes,"* John Wiley & Sons, 2009.

Permissions

All chapters in this book were first published in IJWMN, by AIRCC Publishing Corporation; hereby published with permission under the Creative Commons Attribution License or equivalent. Every chapter published in this book has been scrutinized by our experts. Their significance has been extensively debated. The topics covered herein carry significant findings which will fuel the growth of the discipline. They may even be implemented as practical applications or may be referred to as a beginning point for another development.

The contributors of this book come from diverse backgrounds, making this book a truly international effort. This book will bring forth new frontiers with its revolutionizing research information and detailed analysis of the nascent developments around the world.

We would like to thank all the contributing authors for lending their expertise to make the book truly unique. They have played a crucial role in the development of this book. Without their invaluable contributions this book wouldn't have been possible. They have made vital efforts to compile up to date information on the varied aspects of this subject to make this book a valuable addition to the collection of many professionals and students.

This book was conceptualized with the vision of imparting up-to-date information and advanced data in this field. To ensure the same, a matchless editorial board was set up. Every individual on the board went through rigorous rounds of assessment to prove their worth. After which they invested a large part of their time researching and compiling the most relevant data for our readers.

The editorial board has been involved in producing this book since its inception. They have spent rigorous hours researching and exploring the diverse topics which have resulted in the successful publishing of this book. They have passed on their knowledge of decades through this book. To expedite this challenging task, the publisher supported the team at every step. A small team of assistant editors was also appointed to further simplify the editing procedure and attain best results for the readers.

Apart from the editorial board, the designing team has also invested a significant amount of their time in understanding the subject and creating the most relevant covers. They scrutinized every image to scout for the most suitable representation of the subject and create an appropriate cover for the book.

The publishing team has been an ardent support to the editorial, designing and production team. Their endless efforts to recruit the best for this project, has resulted in the accomplishment of this book. They are a veteran in the field of academics and their pool of knowledge is as vast as their experience in printing. Their expertise and guidance has proved useful at every step. Their uncompromising quality standards have made this book an exceptional effort. Their encouragement from time to time has been an inspiration for everyone.

The publisher and the editorial board hope that this book will prove to be a valuable piece of knowledge for researchers, students, practitioners and scholars across the globe.

List of Contributors

Ming Yan
College of Engineering and Science, VictoriaUniversity, Melbourne, Australia

Hao Shi
College of Engineering and Science, VictoriaUniversity, Melbourne, Australia

Chemseddine BEMMOUSSAT
Dept of Telecommunication, Tlemcen University, Tlemcen, Algeria

Fedoua DIDI
Dept of Computer engineering, Tlemcen University, Tlemcen, Algeria

Mohamed FEHAM
Dept of Telecommunication, Tlemcen University, Tlemcen, Algeria

Soheil Javadi
School of ECE, College of Engineering, University of Tehran, Tehran, Iran

Mohammad H. Hajiesmaili
School of Computer Science, IPM, Tehran, Iran

Behzad Moshiri
School of ECE, College of Engineering, University of Tehran, Tehran, Iran

Ahmad Khonsari
School of Computer Science, IPM, Tehran, Iran

Arvind Kumar
Department of Electronics and Communication Engineering, MNNIT, Allahabad

Rajeev Tripathi
Department of Electronics and Communication Engineering, MNNIT, Allahabad

Emna Trigui
Autonomic Networking Environment, ICD/ERA, CNRS UMR STMR 6279 University of Technology of Troyes, 12, rue Marie Curie, 10010 Troyes Cedex, France

Moez Esseghir
Autonomic Networking Environment, ICD/ERA, CNRS UMR STMR 6279 University of Technology of Troyes, 12, rue Marie Curie, 10010 Troyes Cedex, France

Leila Merghem_Boulahia
Autonomic Networking Environment, ICD/ERA, CNRS UMR STMR 6279 University of Technology of Troyes, 12, rue Marie Curie, 10010 Troyes Cedex, France

A. Daeinabi
Centre for Real-time Information Networks, School of Computing and Communications, Faculty of Engineering and Information Technology, University of Technology Sydney, Sydney, Australia

K. Sandrasegaran
Centre for Real-time Information Networks, School of Computing and Communications, Faculty of Engineering and Information Technology, University of Technology Sydney, Sydney, Australia

X.Zhu
School of Information and Communications, Beijing University of Posts and Telecommunications Beijing, China

Roshni Neogy
Dept. of Information Technology, Jadavpur University

Chandreyee Chowdhury
Dept. of Computer Science & Engineering, Jadavpur University

Sarmistha Neogy
Dept. of Computer Science & Engineering, Jadavpur University

Mohammed Saghir
Hodeidah University, Yemen

Michael Hosein and Laura Bigram
Department of Computing and Information Technology, University of the West Indies, St Augustine, Trinidad

Pantha Ghosal
Centre for Real-time Information Networks School of Computing and Communications, Faculty of Engineering and Information Technology, University of Technology Sydney, Sydney, Australia

Shouman Barua
Centre for Real-time Information Networks School of Computing and Communications, Faculty of Engineering and Information Technology, University of Technology Sydney, Sydney, Australia

List of Contributors

Ramprasad Subramanian
Centre for Real-time Information Networks School of Computing and Communications, Faculty of Engineering and Information Technology, University of Technology Sydney, Sydney, Australia

Shiqi Xing
Centre for Real-time Information Networks School of Computing and Communications, Faculty of Engineering and Information Technology, University of Technology Sydney, Sydney, Australia

Kumbesan Sandrasegaran
Centre for Real-time Information Networks School of Computing and Communications, Faculty of Engineering and Information Technology, University of Technology Sydney, Sydney, Australia

Tripti Sharma
Department of computer science & Engineering, Inderprastha Engineering College, Ghaziabad, (U.P), India

Dr. Vivek Kumar
Department of computer science, Gurukul Kangri Vishwavidyalaya, Haridwar, (U.K), India

Sharmistha Khan
Doctoral Student, Electrical and Computer Engineering Department and ARO CeBCom, Prairie View A&M University, Prairie View, TX, USA

Dr. Dhadesugoor R. Vaman
Texas A&M University System Regents Professor and Director of ARO CeBCom, Electrical and Computer Engineering Department, Prairie View A&M University, Prairie View, TX, USA

Siew T. Koay
Professor, Electrical and Computer Engineering Department, Prairie View A&M University, Prairie View, TX, USA

P. S. Vinayagam
Assistant Professor, Department of Computer Science, Pondicherry University Community College, Puducherry, India

Juhi Garg
Department of Electronics and Communication Engineering, FET-MITS University, Lakshmangarh, Sikar, India

Priyanka Mehta
Department of Electronics and Communication Engineering, FET-MITS University, Lakshmangarh, Sikar, India

Kapil Gupta
Department of Electronics and Communication Engineering, FET-MITS University, Lakshmangarh, Sikar, India

CPSIA information can be obtained
at www.ICGtesting.com
Printed in the USA
BVOW07*2308220816
459704BV00032B/28/P